LAND USE IN AUSTRALIA

PAST, PRESENT AND FUTURE

LAND USE IN
AUSTRALIA

PAST, PRESENT AND FUTURE

EDITED BY RICHARD THACKWAY

Australian
National
University

eVIEW

Published by ANU eView
The Australian National University
Acton ACT 2601, Australia
Email: anupress@anu.edu.au
This title is also available online at press.anu.edu.au

A catalogue record for this
book is available from the
National Library of Australia

ISBN(s): 9781921934414 (print)
          9781921934421 (eBook)

Cover design and layout by ANU Press. Cover photograph: *Sunset* by Ed Dunens, flic.kr/p/qUTrRt.

Contents

Part 1 – The Past and Current Situation

Part 2 – Ad Hoc or Strategic Responses

Part 3 – Working to Achieve National Coordination

Part 4 – Social and Natural Drivers of Change

Part 5 – Visions for the Future

Dedication: Robert George Lesslie

This book is dedicated to the memory of Rob Lesslie, who sadly died in March 2014.

This volume honours Rob's academic contributions and his national influence more generally in land use and management, wilderness protection and the conservation of biodiversity. It is the material legacy of a collaboration that began at the completion of a national land use symposium that was held in June 2015 in his honour, 'Informing Australian land use policy and planning: Past, present and future'.

Rob was a leading Australian geographer and ecologist whose working career spanned more than 30 years in natural resources evaluation and land use and land management in government, education and the private sector. Throughout his career, he published numerous book chapters, journal papers and many technical reports. Rob's impact, however, went well beyond this: he was significantly involved in some of the most pressing natural resource management policies throughout his extensive career.

Rob played a significant role in several initiatives: shaping the Australia China Environment Development Partnership (ACEDP), developing the scientific framework for the WildCountry project and initiating the development and ongoing enhancements and applications of the Multi-Criteria Analysis Shell for Spatial Decision Support. Rob was also the driving force between the establishment of a Memorandum of Understanding between the Australian Bureau of Agricultural and Resource Economics and Sciences (ABARES) and the Chinese Forestry Economics and Development Research Centre to support future cooperative research.

Rob was instrumental in developing and delivering the Australian Government's National Wilderness Inventory program (1986–96) and more recently led an advisory team that developed legislation for wilderness protection in South Australia, the *Wilderness Protection Act 1992*. Currently some 1.8 million hectares of wilderness in South Australia is protected under the Act as a result of a state-wide assessment based on Rob's work.

Over many years, Rob also made major contributions to the classification and mapping of land use, land management regimes and vegetation condition. Information derived from these frameworks and assessments continue to influence decision-makers at national, state and regional levels in the areas of policy formulation, natural resource management and conservation. These endeavours reflect his abiding interest, passion and dedication to the environment—his life's purpose, really.

Rob's generosity and commitment are exemplified through the enduring Lesslie Endowment, which supports research and other activities through The Australian National University (ANU) to promote long-term sustainable management and conservation of Australia's national landscapes and ecosystems. The endowment provides support for research grants, scholarships, fellowships, prizes, public seminars and workshops.

Foreword

Henry Nix AO

The untimely death of Dr Robert 'Rob' Lesslie was the catalyst for the enormous effort that went into the planning and development of this symposium. Rob was a key contributor to both the science and practical application of land use policy and planning, and natural resource management and conservation. While the states and territories have the primary responsibility for sustaining our natural endowment—land, water and life—most of our serious problems transcend these artificial boundaries. A national continental-scale overview then becomes a necessity; here, the Australian Government has a key role in building geographic information systems that inform and support natural resource management. Securing nationwide agreement among all three levels of government was never going to be easy. Rob was not only a good scientist, but also a key member of the negotiating teams that overcame initial interagency suspicion, disagreement, reluctance and opposition. Always enthusiastic, his calm, friendly and professional approach yielded results that were evident in many of the papers presented at the symposium, and in the reports of the panel discussions and plenary conclusions.

Australia is a world leader in the development and application of geographic information systems that provide a foundation for land use policy and planning; however, the take-up in all levels of governance has been erratic and variable. Why is this so? This question provoked a number of responses from contributors and a number of possible explanations and solutions. Inevitably, there are time delays in any organisation in the development and understanding of new technology and its application. Thus, senior management will have to contend with e-infrastructure, high-performance computing, mass-data storage and modelling analysis if we are to move forward.

Land use planning has never been widely popular in democracies, as it is seen to impinge on the rights of the individual. However, a shift from rights to responsibilities is long overdue. There is evidence of some progress at all levels of resource management, ranging from the three levels of governance and corporations to the individual landholder. We do have some exemplars that provide a guide to the new world beyond 'business as usual', but time is short and the need is urgent. Our ability to cope with climate disruption will depend on accurate knowledge of the state and condition of our natural resource base. The papers presented at the symposium made it clear that we have the tools; however, these sometimes yield 'inconvenient truths' that can become political liabilities. Our political systems must find a way to address these issues and not drown them in obfuscation.

The symposium highlighted a wide range of tools and talents and ended on a positive note for the future. However, in March 2016, the Commonwealth Scientific and Industrial Research Organisation (CSIRO) announced a whole-scale demolition of climate, land and water research teams, demonstrating a lack of understanding of the need to monitor the state and condition of our natural resource endowment to achieve sustainable resource management.

Acknowledgements

There are many people to acknowledge in the journey of this book from idea to reality.

In mid-2014, a small group of people touched by the loss of a dear friend and colleague, Rob Lesslie, who sadly died in March 2014, agreed to convene a national symposium in his honour.

The symposium, 'Informing Australian land use policy and planning: Past, present and future', was held in June 2015. It brought together Australia's leading land use researchers, academics and government policymakers and planners. The presenters who were invited all had one thing in common: they had all worked with, or were very familiar with, Rob's academic and professional contributions in land use policy and planning.

The organising committee for the symposium comprised Bruce Doran, Steve Dovers, Jake Gillen and Richard Thackway (Fenner School of Environment & Society, The Australian National University [ANU]); Jason Irving (South Australian Department of Environment, Water and Natural Resources); Brendan Mackey (Griffith University); Jodie Mewett (Australian Bureau of Agricultural and Resource Economics and Sciences [ABARES]). The symposium was held on 29–30 June, in association with the Institute of Australian Geographers conference held on 1–3 July. By aligning the two meetings, the organising committee acknowledged that Rob was also a geographer. Both events were held at the Crawford School, ANU, Canberra.

The authors' contributions to this book honour Rob Lesslie's academic contributions and his national influence more generally in land use and management, wilderness protection and the conservation of biodiversity.

A special thanks to the chapter reviewers. Your work has helped ensure this book is a high-quality publication accessible to policymakers and land use planners, public and private land managers researchers and academic readers.

Australian National University

Griffith UNIVERSITY

ACLUMP
(Australian Collaborative Land Use and Management Program)

INFORMING AUSTRALIAN LAND USE POLICY AND PLANNING: PAST, PRESENT AND FUTURE

Contributors

Valerie A. Brown

Alliance for Regenerative Landscapes and Social Health, Fenner School of Environment and Society, The Australian National University, Canberra, ACT

Val is an Emeritus Professor. She is currently Director, Local Sustainability Project, Fenner School of Environment and Society, The Australian National University. She is also Emeritus Professor of the University of Western Sydney, having been its Foundation Chair of Environmental Health, 1996–2002. Val is the author of several recent books on human capacity for transformational change, tackling wicked problems through the transdisciplinary imagination, collective learning and transformational change, sustainability and health, and social learning and environmental management. In 1999, Val was appointed an Officer of the Order of Australia for research, teaching, policy development and national and international advocacy for sustainable development. She has been appointed a Resident Scholar at the Bellagio Centre of the Rockefeller Foundation.

Brett A. Bryan

Deakin University, Melbourne, VIC

Brett is Professor of Global Change, Environment and Society at Deakin University, Melbourne. His research is focused on creating cost-effective policy for the sustainability of social-ecological systems. As a geographer, Brett has research interests in the application and development of computational tools and analytical methods in a diverse array of social and environmental contexts. Brett's research interests are at the human–environment interface, combining aspects of land use and management; agriculture and food security; water resources management; climate change impact assessment, mitigation, and adaptation; biodiversity conservation; economics and policy analysis.

Tim Clancy

Private Consultant, Brisbane, QLD

Tim was the Director of the Terrestrial Ecosystem Research Network (2011–16), a $67 million NCRIS project delivering critical research infrastructure needed to improve understanding and management of Australia's ecosystems. He worked in senior roles in the Australian Bureau of Agriculture and Resource Economics and Sciences (ABARES) from 2006–11. He was responsible for reporting on national forest, land use, land management and vegetation data. Prior to this, he was Director of the Arthur Rylah Institute for Environmental Research. His other career roles include principal ecologist with NSW State Forests and manager of the Threatened Species Unit for the Queensland Government.

Stephen Dovers

Fenner School of Environment and Society, The Australian National University, Canberra, ACT

Steve was Director of the Fenner School of Environment and Society from 2009 to 2017, and now holds the position of Emeritus Professor. He is an Honorary Professorial Fellow at Charles Darwin University and Fellow of the Academy of Social Sciences Australia. His research and teaching activities relate to the policy and institutional dimensions of natural resources and environmental management, climate adaptation and disasters. His recent works include co-author of the second editions of *Environment and Sustainability: A Policy Handbook* (Federation Press) and *Handbook of Disaster Policies and Institutions* (Routledge).

Mark Eigenraam

Institute for the Development of Environmental-Economic Accounting Group, Fairfield, VIC

Mark is a senior specialist, Environmental Accounting with the Victorian Department of Environment, Land, Water and Planning, and Director at the Institute for the Development of Environmental-Economic Accounting. Over the past 20 years, Mark has played a leading role in the development and application of environmental markets and ecosystem accounting. His work has ranged from training landholders (farmers) to participate in environmental markets, to contributing to global initiatives in environmental-economic accounting. Mark has applied his insights and experience to establishing environmental markets to inform Australia's approach to environmental-economic accounting, and has produced a world-first set of experimental ecosystem accounts for Victoria. He continues to develop new systems and processes to produce and publish environmental-economic accounts.

Siddeswara M. Guru

University of Queensland, St Lucia, Brisbane, QLD

Siddeswara is a data integration and synthesis manager for the Terrestrial Ecosystem Research Network. He initiates, coordinates and manages ecological data, e-infrastructure and synthesis projects; he also oversees the data and information management activities across the Terrestrial Ecosystem Research Network (TERN). Siddeswara has a strong research and management background. He was awarded a PhD from the University of Melbourne and an MBA from the University of Tasmania. Previously, he worked as a data scientist and project officer at the Integrated Marine Observing System and data management officer at CSIRO (Marine and Atmospheric research). He held a post-doctoral fellowship at CSIRO's Tasmanian ICT Centre, where he worked on environmental sensor data management.

John A. Harris

Alliance for Regenerative Landscapes and Social Health, Fenner School of Environment and Society, The Australian National University, Canberra, ACT

John has wide experience as a researcher, teacher and community activist in the fields of ecology, conservation and environmental education, with emphasis on the connections between theory and practice. His academic appointments include the University of Canberra, The Australian National University, CSIRO, Colorado State University and the University of Hanoi. He is co-author and co-editor, with Valerie A. Brown, of *The Human Capacity for Transformational Change: Harnessing the Collective Mind* and *Tackling Wicked Problems Through the Transdisciplinary Imagination*, and author of *The Change Makers: Stories from Australia's First Environmental Studies Graduates*.

Richard W. Hicks

Private Consultant, Dubbo, NSW

Richard was previously the Chair of Australian Collaborative Land Use and Management Program (ACLUMP) and the NSW representative on the committee. He has an extensive background in remote sensing and land use planning, focusing on regional and fine-scale local government mapping, and was the leader of remote sensing and land assessment programs for the NSW Office of Environment and Heritage prior to retiring in 2017. Richard was instrumental in developing the NSW remote-sensing satellite-monitoring program and the associated land use planning activities. He was the NSW lead in setting up the Joint Remote Sensing Research Program with the University of Queensland and the Queensland, NSW and Victorian governments.

Gary Howling

Office of Environment and Heritage, Wollongong, NSW

Gary is currently conservation manager with the Great Eastern Ranges Initiative. He is responsible for providing regional and program-wide partnerships with specialist scientific conservation advice to guide the development of connectivity conservation projects. His role also involves communicating the importance of collaboration across and between stakeholders and landscapes.

Jason Irving

Department of Environment, Water and Natural Resources, Adelaide, SA

Jason is the manager of the Protected Areas Branch, Department of Environment, Water and Natural Resources, South Australia. He is responsible for legislation and policy for protected areas on public and private land, including developing policy to protect Arkaroola in the Flinders Ranges through the *Arkaroola Protection Act*. He worked with the Australian Committee for the International Union for the Conservation of Nature (IUCN) on the publication *Innovation for 21st Century Conservation*. In addition, he has overseen the development of co-management arrangements for parks with traditional owners, prepared the state protected area strategy and overseen the proclamation of over 100 new parks or additions to parks.

Christina Jones

Department of Environment and Science, Brisbane, QLD

Christina is currently the manager of the Remote Sensing Centre in the Queensland Department of Environment and Science. She is responsible for state-wide landscape monitoring and mapping programs, including vegetation change and land use mapping. The centre contributes to the Joint Remote Sensing Research Program, a collaborative science partnership between the University of Queensland, the University of NSW, and the Queensland, NSW and Victorian governments.

Valdis Juskevics

Australian Bureau of Statistics (retired)

Valdis had over 40 years' experience as a social, economic and environmental statistician with the Australian Bureau of Statistics. Before retiring, he undertook a three-year secondment to Geoscience Australia to assist in the work associated with the development of the Geoscience Australia's National Exposure Information System, a tool that provides nationally consistent aggregated (disaster) exposure information irrespective of existing administrative or geographic boundaries. He co-authored a significant study into the impact of the 2011 and 2013 Brisbane and Ipswich floods on households.

Paul Lawrence

Department of Environment and Science, Brisbane, QLD

Paul is the Director for Landscape Sciences in the Queensland Government. He has worked in natural resource management for 35 years, undertaking monitoring, modelling and providing science to inform policy and planning. He has Bachelor and Masters degrees from Griffith University and a PhD from the University of Arizona. He completed an Organisation for Economic Co-operation and Development (OECD) postdoctoral fellowship and an Executive Masters in Public Administration from Monash University. He is the Queensland representative on the National Soils Network for Research Development & Extension, and was chair of the National Committee on Land Use and Management Information.

Rob Lesslie

Fenner School of Environment and Society, The Australian National University, Canberra, ACT

Rob was a geographer and ecologist with interests in natural resources evaluation, land use and management in government, education and the private sector. He was highly influential in developing and delivering the Australian Government's Australian Collaborative Land Use and Management Program (ACLUMP), which promotes collaboration among Australian and state government agencies and others with interests in land use change analysis. More recently, Rob led an advisory team that developed legislation for wilderness protection in South Australia. Rob also developed the National Wilderness Inventory program (1986–96). Sadly, Rob passed away on 28 March 2014. The production of this book commemorates his leadership in the areas of land use policy and planning.

Darryl Low Choy

Griffith University, Nathan, QLD

Darryl is Professor of Environmental and Landscape Planning at Griffith University. He is leading research into climate change adaptation for human settlements and resilient communities' responses to, and recovery from, natural hazards. He is a member of the Cooperative Research Centre for Water Sensitive Cities, which is researching catchment-scale landscape planning for water-sensitive cities in an age of climate change. He is a Registered Planner and Fellow of the Planning Institute of Australia and has extensive industry experience. He has completed several major secondments to state government planning initiatives. Darryl has a Visiting Professorship for Senior International Scientists of the Chinese Academy of Sciences and, in 2016, was awarded a Chinese Academy of Sciences President's International Fellowship.

Brendan Mackey

Griffith University, Gold Coast campus, QLD

Brendan is Director of Griffith University's Climate Change Response program. He serves on the Council of the IUCN and chairs its climate change task force. He is also a member of the Great Eastern Ranges Science Panel. Brendan has over 150 publications in the fields of biogeography, conservation, climate change and science-based land use policy.

David Marsh

Alliance for Regenerative Landscapes and Social Health, Fenner School of Environment and Society, The Australian National University, Canberra, ACT

David has been farming for 45 years at Boorowa, NSW. He has been involved in conservation, both on his farm and in various organisations. For the past 17 years, his family has been managing the farm holistically—that is, making decisions that are socially, economically and environmentally sound. He won the Central West Conservation Farmer of the Year award in 2004. He graduated with a Masters in Sustainable Agriculture from the University of Sydney in 2001. The Marshes have been managing in a way that allows the regenerative capacity of earth to become a reality. This style of management leads to increasing complexity. David has given talks on regenerative agriculture and managing holistically in all states of Australia (except the Northern Territory). He has been a member of the Native Vegetation Advisory Council of NSW, the Lachlan Catchment Management Authority and Soils for Life. He is a founding member of the Alliance for Regenerative Landscapes and Social Health and writes a blog on their website.

Neil McKenzie

CSIRO Land and Water, Canberra, ACT

Neil has more than 30 years' research experience in the land and water sciences. His work has focused on quantitative methods for mapping soil and land resources. Neil's team was responsible for several national standards on measurement, monitoring and survey. He was Chief of CSIRO Land and Water from 2007–12 and continues to be involved in shaping policy on scientific aspects of land resource management in Australia and internationally. Neil is currently a member of the Intergovernmental Technical Panel on Soils and leads several projects that aim to improve soil management in Australia and the Pacific.

Jodie Mewett

Australian Bureau of Agricultural and Resource Economics and Sciences, Canberra, ACT

Jodie is a scientist in the Australian Bureau of Agricultural and Resource Economics and Sciences (ABARES). She has 13 years' experience in collating, analysing, promoting and providing advice on land use and land management practices information. She is the coordinator of the Australian Collaborative and Land Use Management Program, a consortium of Australian and state government partners. She is an expert in the Multi-Criteria Analysis Shell for Spatial Decision Support (MCAS-S) tool and has applied MCAS-S in a variety of national and international projects.

John Neldner

Department of Environment and Science, Brisbane, QLD

John manages a team involved in ecological research, focusing on the dynamics and condition of ecosystems across Queensland. John has extensive vegetation survey and mapping experience, and has chaired Queensland's Species Technical Committee, which assesses species for conservation status under the *Nature Conservation Act*, for the past seven years.

Henry Nix

Fenner School of Environment and Society, The Australian National University, Canberra, ACT (retired)

Henry was Professor at the Centre for Resource and Environmental Studies at The Australian National University between 1986 and 2002, and its Director for 14 years. He was Emeritus Professor and visiting fellow at the Fenner School of Environment and Society from 2002–10. His research interests include macro-ecology, simulation of agricultural and biological systems, prediction of plant and animal distributions, environmental history, climatology and ornithology. He has received numerous awards including the Gold Medal of the Ecological Society of Australia in 1994; in 2000, he was appointed an Officer of the Order of Australia.

Phillip Norman

Department of Environment and Science, Brisbane, QLD

Phil is a principal scientist with the Queensland Department of Environment and Science. He is based at the Ecosciences Precinct in Brisbane. He is currently working on assessing biomass feedstocks across Queensland. Previously, Phil provided scientific input and oversight to the Queensland Agricultural Audit, Queensland State Planning Policy (with respect to agricultural land use) and the South East Queensland

Forest Assessment. Phil's research interests are focused around land planning, sustainable use of land and vegetation, and ways and means of reconciling competing demands for land resources.

John Ovington

Australian Bureau of Statistics (retired)

John was Assistant Director at the Australian Bureau of Statistics (ABS) Centre of Environment and Energy Statistics, where he was responsible for a range of environmental statistics, including water, energy and land. He has worked on a variety of other ABS surveys, including household surveys and industry statistics. John is now retired and has more time to travel and fish.

Ian Pulsford

Private Consultant, Canberra, ACT

Ian has over 36 years' experience in protected area selection, design and management, conservation planning and connectivity conservation. He has published many papers and articles and co-edited two books on linking Australia's landscapes and protected area governance and management. During the 1990s and later, Ian was the National Parks and Wildlife Service Zone and Divisional Manager for Conservation Programs in south-east NSW. From 2007 to 2010, he was the founding manager of the Great Eastern Ranges Initiative. Ian is a member of the IUCN World Commission on Protected Areas – Connectivity Conservation Specialist Group, Southern National Parks Advisory Committee and Great Eastern Ranges Board.

John Purcell

Australian Bureau of Statistics (retired)

John has spent over 15 years working on developing and implementing environmental statistical and accounting solutions and information systems for governments at the national, state and regional level. John is self-taught in statistics and mathematics and has worked for over 30 years at both the Australian Bureau of Statistics and Murray–Darling Basin Authority. John is now retired.

Jacki Schirmer

University of Canberra, ACT

Jacki is an Associate Professor with the Centre for Research and Action in Public Health, University of Canberra Health Research Institute. She has examined the social dimensions of natural resource management for 15 years. Her work focuses on understanding the relationship between human wellbeing and environmental change, particularly changes in land and water use. She has a particular interest in community engagement, conflict resolution and adoption of new conservation practices.

Craig Shephard

Department of Environment and Science, Brisbane, QLD

Craig is the principal scientist in the Queensland Land Use Mapping Program within the Department of Environment and Science, Queensland Government. He has 20 years' professional experience in the mapping of landscape attributes, particularly land cover and land use in Queensland. Previously, he spent two years mapping crime in London. Craig holds a Bachelor of Science. He is interested in the application of geographic information system technology to natural resource management.

Richard Thackway

Fenner School of Environment and Society, The Australian National University, Canberra, ACT

Richard is an Adjunct Fellow in the Fenner School of Environment and Society, The Australian National University, an Adjunct Associate Professor in School of Geography, Planning and Environmental Management, University of Queensland, and a Visiting Fellow at the UNSW Australian Defence Force Academy. His research aims to help others improve their decision-making about natural resources, by developing and implementing spatial and temporal decision-support tools, frameworks and information systems for assessing and reporting natural resource condition associated with land use and land management.

Michael Vardon

Fenner School of Environment and Society, The Australian National University, Canberra, ACT

Michael is an expert in environmental accounting and has been collecting and analysing data for more than 20 years. He is currently researching and teaching environmental accounting at The Australian National University. He is a member of the Technical Expert Committee of the World Bank's Wealth Accounting and Valuation of Ecosystem program, and was a member of the editorial board of the System of Environmental-Economic Accounting. He was formerly the Director of the Centre of Environmental Accounting, a position he held from 2005 until 2014, with secondments to the United Nations 2007–09 and Bureau of Meteorology 2012–13.

Christian Witte

Department of Environment and Science, Brisbane, QLD

Christian is the Director for Environmental Monitoring and Assessment Sciences. This includes monitoring programs for water quality, air quality, wetlands and biological surveys, as well as a range of environmental investigations. Christian was previously responsible for the Remote Sensing Centre and established the Queensland Land Use Mapping Program. He has completed a science degree and a Masters in Geographic Information Systems (GIS) at the University of Queensland and a Graduate Certificate in Change Management at the Australian Graduate School of Management, University of NSW.

Glossary of Terms

ABARES	Australian Bureau of Agricultural and Resource Economics and Sciences
ABS	Australian Bureau of Statistics
ACEDP	Australia China Environment Development Partnership
ACLEP	Australian Collaborative Land Evaluation Program
ACLUMP	Australian Collaborative Land Use and Management Program
AEEA	Australian Environmental-Economic Accounts
ALA	Atlas of Living Australia
ALUM	Australian Land Use and Management
ALUMC	Australian Land Use and Management Classification
AML	Arc Macro Language
ANWI	Australian National Wilderness Inventory
ANZECC	Australian and New Zealand Environment and Conservation Council
ArcGIS	Earth Sciences and Resources Institute's platform that enables its GIS users to discover, use, make, and share maps
ArcSDE	Earth Sciences and Resources Institute's server-software sub-system, i.e. Spatial Database Engine
AVHRR	Advanced Very High Resolution Radiometer
AWRAP	Australian Water Resources Assessment Programme
BoM	Bureau of Meteorology
BRS	Bureau of Rural Sciences

Cadastre	A comprehensive register of the real estate or real property's metes and bounds of a country. The cadastre is a fundamental source of data in disputes and lawsuits between landowners
CAWCR	Collaboration for Australian Weather and Climate Research
CoESRA	Collaborative Environment for Ecosystem Science Research and Analysis
Copernicus	Satellite program of the European Space Agency covering land surface, marine and meteorological satellites
CPU	central processing unit
CSIRO	Commonwealth Scientific and Industrial Research Organisation
DAFF	Department of Agriculture, Fisheries and Forestry
DCDB	Digital Cadastral Database
DPE	Departments of Planning and Environment
DSITI	Department of Science, Information Technology and Innovation
EBI	Environment Benefit Index
Econd	Environment condition index
EEA	Experimental Ecosystem Accounting
EPBC	*Environment Protection and Biodiversity Conservation Act 1999*
ETM+	Enhanced Thematic Mapper Plus
EU	European Union
GDP	gross domestic product
GER	Great Eastern Ranges Initiative
GHG	greenhouse gas emissions
GIS	geographic information system
GPU	graphics processing unit
HPC	high-performance computing
ICT	information communication technology
IT	information technology

ITPS	Intergovernmental Technical Panel on Soils
IUCN	International Union for the Conservation of Nature
IVG	Independent Verification Group
JANIS	Joint ANZECC – MCFFA National Forest Policy Statement Implementation Subcommittee
LUMIS	Land Use Management Information System
MCAS-S	Multi-Criteria Analysis Shell for Spatial Decision Support
MCFFA	Ministerial Council on Forestry, Fisheries and Aquaculture
MDBA	Murray Darling Basin Authority
MODIS	Moderate Resolution Imaging Spectroradiometer
MODIS EVI	MODIS Enhanced Vegetation Index
MODIS NBAR	MODIS Nadir Bi-directional Reflectance Distribution Function Adjusted Reflectance
NASA	National Aeronautics and Space Administration
NCA	*Nature Conservation Act 1992*
NCLUMI	National Committee on Land Use and Management Information
NCRIS	National Collaborative Research Infrastructure Strategy
NCST	National Committee on Soil and Terrain
NLWRA	National Land and Water Resource Audit
NRM	natural resource management
NRMMC	Natural Resource Management Ministerial Council
OECD	Organisation for Economic Co-operation and Development
PUCE	Pattern-Unit-Component-Evaluation program
QLUMP	Queensland Land Use Mapping Program
QRISCloud	Queensland Research and Innovation Services Cloud
QVAS	Queensland Valuation System
RAC	Resource Assessment Commission

RFA	Regional Forest Agreement
SEEA	System of Environmental-Economic Accounting
Sentinel	Land surface satellites within the Copernicus Program, the Sentinel 2 satellite is a 10-metre resolution platform offering similar wavelength detection to that of Landsat 8
SEQ	South East Queensland
SIAP	Spatial Imagery Acquisition Program
SPOT	A commercial constellation of satellites comprised of a high-resolution optical imaging Earth observation satellite system. It is run by Spot Image, based in France.
TERN	Terrestrial Ecosystem Research Network
TFA	Tasmanian Forest Agreement
TM	Thematic Mapper
UHSRLUP	Upper Hunter Strategic Regional Land Use Plan
UN	United Nations
UNFCCC	United Nations Framework Convention on Climate Change
VAST	vegetation assets, states and transitions
VMA	*Vegetation Management Act 1999*
WAC	Wilderness Advisory Committee
WMS	Web Map Services
WONS	weeds of national significance
WWII	World War II

Introduction

Richard Thackway

Australia's landscapes support a wide range of land use activities, including intensive agriculture, pastoralism, nature and heritage conservation, and forestry and defence training. In addition, Australia's landscapes have been under Indigenous land management for at least 50,000 years. These land use activities occur across a mix of public, private, leasehold and Indigenous land tenures, utilising native vegetation that ranges in condition from variously unmodified to modified and replaced (Thackway & Lesslie, 2008). In 2012, production from natural resources earned over $38 billion in exports from agriculture, fisheries and forestry (see Chapter 3). The total economic value of Australia's land use activities, inclusive of ecosystem services and non-consumptive uses, remains to be estimated.

The way in which land is used has a profound effect on Australia's unique climate, soil, water, vegetation and biodiversity resources (Thackway & Freudenberger, 2016). There is a strong link between spatial and temporal patterns of land use and prevailing environmental, economic and social conditions. Therefore, information on land use and management is fundamental to the development and implementation of land use policy and planning.

Planned land use choices have a major effect on our natural environment, our communities and the capacity of regions to produce food and maintain and protect biodiversity and ecosystem services. Land use policies and planning are central to debates in Australia around coal and gas extraction, urban expansion, water security, climate change adaptation, population and food security. Under the influence of a rapidly changing climate, informed land use policies and planning are critical to developing effective responses to natural resource management imperatives, including biodiversity protection, water quality, water and food security, as well as sustainable production from agricultural areas, forests and rangelands.

This book presents critical insights into successes, failings and solutions from land use policy and planning case studies. It seeks to address four critical issues facing Australia:

1. discontinuities between providers of national biophysical information about the environment, agriculture and forestry, among other land use activities

2. poor awareness of tools and information to improve national land use decision-making outcomes

3. limited awareness of the long history of poor decision-making on land use planning and management across different levels of government

4. limited understanding of the benefit of land use practitioners engaging in mutual inquiry and collective learning by working with key decision-makers from a range of fields, including environment, agriculture, health, nutrition, planning, Indigenous interests, management, design, education and research.

Much more can be achieved. The findings presented in this book strategically highlight the benefits of tracking changes and trends in the extent of land use types at different spatial scales over time; opportunities for improving monitoring and evaluation of social, environmental and economic outcomes arising from land use decisions made at multiple scales; prospects for utilising cutting-edge tools and information, including imagery archives and research to support decision-makers; and the benefits of using continual social learning to establish and develop land use policies, and evaluate the outcomes of implementing land use policies and plans.

Barriers to the Adoption of New Ideas

Successive governments have invented and reinvented 'fad'-based land use policies and plans, reflecting short-term thinking and partisan ideals and solutions to the long-running, recurring and pressing land use policy and planning issues that face communities at regional, state and national levels (see Chapter 4). Recognising the weakness of this short-term populist approach to policy and planning, some leading Organisation for Economic Cooperation and Development (OECD) member governments have expressed a desire for 'fact' or evidence-based approaches (Morton & Tinney, 2012). Where land planning and management agencies are committed to practising evidence-based approaches to developing and resolving national policy and program issues, they can develop robust and enduring systems with standards for the collection, analysis and provision of (and access to) high-quality natural resources and social and economic information. Valuing collective community ideals and supporting access to relevant information is vital if we are to overcome the current malaise in land use policy and planning. Not addressing (and overcoming) the ongoing reluctance to engage communities of interest perpetuates adversarial approaches to establishing, developing and implementing land use policy and planning. This book presents case studies that demonstrate the shortcomings of top-down government-led approaches to developing sustainable futures (see Chapters 4 and 5), including the challenges of anthropogenic climate change (see Chapter 15).

One of the principles for improved social, cultural and environmental outcomes is the provision of opportunities to establish and develop coordinated approaches to enable continuous participation of interested communities in the policy and planning processes; another principle is to develop institutional arrangements to effectively use relevant and up-to-date information, including tools to aid in decision-making. Neither of these principles are being effectively enabled by government. Instead, as shown across several of the case studies presented here, they are either being actively diminished as not relevant, filled by non-government agencies and industry bodies, or divested to the states and territories in ways that do not enable the necessary cross-jurisdictional harmonisation and national-level coordination (see Chapters 11, 12 and 17).

Arguably, solutions to our land use problems do not lie only in accessing better data and information. We are experiencing a data explosion as access to near real-time environmental information improves, motivating the development of new land management techniques. A key problem lies in our democratic and governance structures, which do not seek the collective participation of decision-makers who aim to develop common property resources, or interested communities that will be affected by short-term and populist approaches to policy and planning processes (Brown & Harris, 2014). This process is exacerbated when the emphasis of policy and planning is weighted in favour of short-term, narrowly defined economic development, rather than environmental, social and economic indicators. Unless these major barriers towards adopting balanced and sustainable solutions are addressed, we will continue to see limited improvement in land use policy and planning initiatives.

The development and implementation of efficient and effective national land use policy and planning are hindered by a range of social, cultural and environmental barriers. These barriers are neither new, nor restricted to land use policy and planning. In a review of the processes for the investment in (and management and use of) environmental information as part of the initiative for a National Plan for Environmental Information, Morton and Tinney (2012) identified opportunities for improving the efficiency and effectiveness of the Australian Government's environmental information. Given the enormity of the barriers facing land use policy and planning in Australia, new governance arrangements should be established to ensure that policy priorities and strategic directions are set at the whole-of-government level. Several of the case studies presented here highlight that land use policy and planning are in disarray, and lack focus, determination, coordination and leadership (see Chapters 5, 9, 10 and 15).

The monitoring and evaluation of land use policy and planning outcomes are implemented neither consistently nor comprehensively across all jurisdictions. Where national monitoring is undertaken, it tends to focus narrowly on a specific theme—for example, carbon, biodiversity or water. Such monitoring initiatives show what is technically possible when there is a strong link between policy and political imperatives. However, the lack of comprehensive monitoring hinders the ability of regional, state and national public–private agencies to review and adjust land use policy settings and program outcomes (Sbrocchi et al., 2015). Consequently, the capacity to develop balanced outcomes for Australian landscapes across the spectrum of natural resource management issues—biodiversity, carbon, agricultural productivity, biosecurity, water security and food security—is diminished (see Chapter 3). Several theme-based examples of monitoring are presented here (see Chapters 8 and 13). However, as useful as these are, broad-based national environmental accounts are needed to inform decision-makers of changes and trends in the natural resource base. When these accounts are integrated with economic and social information, key insights begin to emerge to inform land use policymakers and planners (see Chapter 14).

Drivers of Land Use Change

Some land use change can involve major social, environmental and economic turmoil, resulting in the need for adjustment—for example, water reforms in the Murray–Darling Basin, moratoriums on tree clearing, the cessation of logging in native rainforests, coal gas extraction in prime agricultural land and urban development in ecologically sensitive areas. The drivers of land use change are complex and subject to change over various spatial and temporal scales. They include:

- government policy and program interventions, such as access to resources, regulatory and legislative opportunities and constraints, governance arrangements, various instruments, and incentives and disincentives (e.g. infrastructure, subsidies and taxes)
- new commodities and land use activities
- market access, transport costs, new technologies, infrastructure and production costs
- changing societal needs for access to land for different purposes, and the skill and knowledge levels required to manage land with varying degrees of suitability and capability (while maintaining or transitioning its resource condition)
- the social resilience of regional communities to learn and adapt to complex interactions amid cultural and biophysical changes, and the capacity to handle the natural variability of climate and climate change, managing enterprises in the face of various natural hazards and disasters
- anthropogenic climate change.

Over the last 50 years, Australia's population in rural areas and the proportion of Australia's land area used for agriculture have slowly declined—a trend that can be described as 'extensification'. Set against this trend is 'intensification', which is the movement of people towards the coastal hinterlands and major cities located in southern and eastern Australia. The increasing use of mechanisation in broadacre agriculture and forestry production and the conversion of prime agriculture land for rural–residential and residential development are associated with intensification.

Opportunities to enable greater and more regular community engagement, and influence these land use policy and planning drivers at regional, state and national levels, are currently limited because of the reliance on top-down government-led and controlled initiatives (Brown & Lambert, 2013).

Book Structure

This book aims to fill a gap in current understandings of the barriers to good land use policy and planning in Australia by showing the contribution that information systems and decision support tools can make to the sustainable management of our unique landscapes. The authors, all highly regarded land use researchers, policymakers and practitioners, draw on their experiences to present short and accessible case studies on previous and current work relating to land use policy and planning. As well as providing deep insight into Australian land use issues, these case studies demonstrate the information and tools currently available to improve policy and decision-making at regional, state and national levels.

Part 1 examines the *past and current situation*. Comprising four chapters, it sets out the status of land use and management, as well as the challenges facing land use policy and planning in Australia. The role of Rob Lesslie—a highly influential land use planner, landscape scientist, geographer and wilderness specialist—highlights the contribution that key individuals can make in initiating and leading change in land use policy and planning in Australia. This section acknowledges that public policy is driven by social, political and economic factors other than data and information relevant to land use, and poses the rhetorical question: does public policy obey data, information and maps?

Part 2 presents a series of case studies under the heading *ad hoc or strategic responses*. Across four chapters, it examines relationships and issues that have arisen between land use planning initiatives and environmental (landscape) resource mapping initiatives, processes for identifying and responding to land use pressures at the state and territory level, approaches for balancing land use trade-offs, the role of wilderness in nature conservation in South Australia and the effects of land use change on biodiversity in Australia.

Part 3 presents two chapters that address the theme *working to achieve national coordination.* Given that land use policy and planning are state and territory responsibilities (under the Australian Constitution), and that national land use policy and planning can only be achieved through national leadership to achieve national coordination with state and territory agencies, these two chapters focus on the need for national coordination of vegetation and soil matters.

The *social and natural drivers of change* are explored in Part 4. Five chapters present case studies covering a wide range of issues: environmental conflict, engaging with scientific information and community activism, landscape connectivity initiatives, approaches to monitoring and reporting land use change and its effects on the environment, the role that national environmental accounts can play in shaping land use policy and planning, and the challenge of a rapidly changing climate in national land use policy and planning.

Two chapters offering *visions for the future* are presented in Part 5. Two key integration issues are discussed: the need for better national e-infrastructure, modelling analytics and synthesis, and securing institutional capacity for land use policy and planning and opportunities; and the benefits of the 'collective learning spiral' in the case of regenerative landscape policy and practice. Both chapters offer possible pathways to better inform national land use policy and planning.

References

Brown, V. A. & Harris, J. A. (2014). *The human capacity for transformational change: Harnessing the collective mind.* London, UK: Routledge.

Brown, V. A. & Lambert, J. A. (2013). *Collective learning for transformational change: A guide to collaborative action.* London, UK: Routledge.

Morton, S. & Tinney, T. (2012). *Independent review of Australian Government environmental information activity.* Retrieved from www.environment.gov.au/system/files/pages/8d3f2610-c336-4e47-aaa7-f3d2b879b905/files/eia-review-discussion-paper.pdf

Sbrocchi, C., Davis, R., Grundy, M., Harding, R., Hillman, T., Mount, R., ... Cosier, P. (2015). *Evaluation of the Australian regional environmental accounts trial.* Sydney, NSW: Wentworth Group of Concerned Scientists.

Thackway, R. & Freudenberger, D. (2016). Accounting for the drivers that degrade and restore landscape functions in Australia. *Land 5*(4), 40. doi.org/10.3390/land5040040

Thackway, R. & Lesslie, R. (2008). Describing and mapping human-induced vegetation change in the Australian landscape. *Environmental Management 42*, 572–90. doi.org/10.1007/s00267-008-9131-5

Part 1 – The Past and Current Situation

1

Australian Land Use Policy and Planning: The Challenges

Richard Thackway

Key Points

- Land use decisions made during the early years of European settlement are evident in today's broad land use patterns; the location of major population centres; issues of land degradation; patterns of remnant vegetation; and responses to vulnerable, threatened and rare species. This creates issues and challenges that require national coordination and the development of land use strategies.

- In areas that have been cleared for agriculture (primarily cereal cropping) and grazing on improved pastures, remnant areas of native vegetation provide a much-reduced, but vitally important, reservoir for biodiversity.

- In cleared landscapes, issues of soil salinity and acidification create ongoing challenges for governments and industry bodies faced with developing agricultural land use policies and sustainable management systems.

- To avoid perverse planning and land use outcomes, policymakers and land planners need access to nationally consistent, high-quality and up-to-date scientific information and scenario tools.

Australia's Land Use Policy and Planning

Australia comprises a land area of almost 7.7 million square kilometres, and was settled and developed between 1788 and 1950—a process that displaced its Indigenous peoples. The contrast between European culture and Indigenous cultures resulted in major ecological and social transformations across the continent, while the establishment of settled agricultural practices resulted in major transformations of the landscape and the fragmentation and modification of native vegetation (Thackway & Lesslie, 2008). Australia has experienced the world's highest rate of extinctions of mammalian fauna, and parallel losses of biodiversity across many of its ecosystems (Steffen et al., 2009).

Living with the threat of droughts, floods and wildfires of devastating proportions, land managers and developers of land use policy and plans have adapted and learned new systems of agriculture over time. For those from more predictable European origins, learning to understand and adapt to new-world patterns of long-term climate variability was difficult. Arguably, today's decision-makers are generally more aware of, and attuned to, El Niño and La Niña events, which are a natural part of the global climate system, than previously. Across our agro-climatic regions, La Niña events are often associated with higher-than-average rainfall, and are triggers for bumper crops and outbreaks of pest animals and plants— the antecedents for fuel build-up and major wildfires. El Niño events are associated with prolonged periods of below-average rainfall and, at times, devastating droughts. The rapid onset of a La Niña event, following a severe El Niño event, has been associated with major soil erosion events.

Historically, issues of low soil fertility, and low and highly variable rainfall across most of the hinterland, presented major challenges for governments and land managers, as these problems were unknown in European experience. Land use decisions made during the early years of European settlement continue to reverberate today. They are evident in land use patterns; the location of major population centres; issues of land degradation; patterns of remnant vegetation; and responses to vulnerable, threatened and rare species.

Australia is a federation of eight states and territories. Under the Australian Constitution, each jurisdiction has responsibility for land use policy and planning. As with many areas of national natural resource management (NRM), including land use policy and planning, the Australian Government plays a major role or takes the lead in developing partnerships between key national, state and regional stakeholders; establishing and fostering the adoption of national standards and protocols for collecting consistent national datasets; compiling data and information for national and international monitoring and reporting; coordinating cross-border NRM initiatives at a range of spatial and temporal scales; and fostering participation in the development of cross-border land use policies and planning to improve decision-making across space and time.

Almost 64 per cent of Australia's land is being used to meet domestic food and fibre needs, and to provide export income. Less than 4 per cent of Australia's total land area is under intensive agricultural and urban use; in general, native vegetation has been completely removed from such areas. About 2 per cent of the country is used for production forestry, including plantations. Around 57 per cent of the land area has been modified for agricultural and pastoral use, with the bulk of this area being used for livestock grazing on natural pastures. Only 10 per cent of Australia's landmass has been converted from native vegetation to intensive agricultural production, including modified pastures. In areas that have been cleared for agriculture (primarily cereal cropping) and grazing on improved pastures, remnant areas of native vegetation provide a much-reduced, but vitally important, reservoir for biodiversity. In these cleared landscapes, issues of soil salinity and acidification create ongoing challenges for governments and industry bodies faced with developing agricultural land use policies and sustainable management systems.

In 2015, Australia's population was almost 24 million. Five major state capitals, each with a population greater than 1 million, were home to more than 14 million people. These cities—Perth, Sydney, Melbourne, Brisbane and Adelaide—are located in the coastal margin of the south-west, southern and eastern states. Only around 2 million people occupy Australia's interior—much of which is semi-arid and arid rangelands. Such pronounced clustering of people in the coastal zone and immediate hinterland, and across several jurisdictions, creates issues and challenges that require national coordination and the development of land use strategies.

The population of Australia has doubled since 1970. Since the 1950s, Australia's population has transitioned from rural communities to urban and peri-urban dwellers. The bulk of Australia's population is distributed predominantly within 100 kilometres of its coastline. This growth has placed pressure on the coastal zone and its hinterland to provide more land for urban and peri-urban dwellers. Increasing population growth and the pressure to provide access to land for urban and peri-urban development in the coastal zone are perennial problems for decision-makers involved in land use policy and planning. Issues for planning include the conversion of agricultural land use to urban and peri-urban areas. Along with the demand for suitable urban land, there are demands to plan for and provide more open space in urban areas to make cities more liveable. Additionally, a steady demand for access to potable water, clean air, nature-based tourism and conservation reserves (inside and outside) for the protection of biodiversity has accompanied the burgeoning growth of urban and semi-urban areas. To avoid perverse outcomes, policymakers and land planners need access to nationally consistent, high-quality and up-to-date scientific information and scenario tools.

Historically, Australia's economy has been based on developing, servicing and exporting its natural resources, including agricultural and forest products and minerals. In some cases, developing these resources has seen major tension and conflict between those in favour of development and those seeking conservation and protection of natural landscapes. Essentially, these issues involve trade-offs between various aesthetic, social, economic and environmental values. Land use policies and planning transform landscapes, affecting the long-term mosaics of unmodified, modified, removed and replaced vegetation ecosystems. In turn, land use policies, planning instruments and decisions affect the viability of landscapes to generate publicly acceptable mixes of ecosystem services, including clean air, healthy crops, clean water, and parks and reserves for the protection of nature and recreation. Again, such key issues require coordinated national land use policies and initiatives.

In 2015, nearly 37 per cent of Australia's land was in national parks, nature reserves and other protected areas. Such areas, in which the effects of land use on the environment are limited or negligible, contribute to biodiversity conservation objectives. However, despite being little affected by (or in advanced recovery from) clearing, the impact of fire and feral animals can be significant. A commitment by all Australian governments, including states and territories, over almost 20 years is responsible for the relatively large area designated for reserve and off-reserve protection of biodiversity. Regrettably, this coordinated national approach largely ceased in 2010, when the Australian Government withdrew resources from the process.

In 2015, Australia was recognised as having one of the highest gross domestic products in the world. In terms of its natural landscapes and clean and green food production systems, it was also recognised as one of the most liveable countries. However, this development has come at a cost, as Australia also has one of the highest carbon dioxide emissions per head of population of any nation.

While Australia's world ranking for natural landscapes, clean and green food production systems and standard of living is high, there are serious problems at the national level in terms of land use policy, including:

- poor coordination and support for NRM, conservation and land use planning
- limited active and current awareness of recent land use decisions, resulting in unforeseen consequences and poor NRM outcomes
- poor awareness and use of nationally consistent, well-maintained information systems developed to support policy and planning at the multiple scales at which decisions are made—that is, national, state, regional, local and non-spatial decision loci, such as issue or sectoral (these information systems are not widely used: why?).

This book is the product of a symposium on land use policy and planning that was held at the Crawford School of Public Policy, The Australian National University, in June 2015. The aims of the symposium were twofold: to review advances in approaches to assessing and mapping land use, ecological condition and related ecosystem services; and to consider how this information could be better generated, managed and used to improve policy and decision-making. Given the magnitude of these aims, we decided to bring key decision-makers in land use policy and planning together with academic researchers and practitioners. All speakers were leaders in their field; they were selected for their capacity to inform and improve the quality of national land use planning and to limit the effects of perverse outcomes in the following areas:

- conversion of prime agricultural land to urban and mining
- loss of biodiversity and natural heritage values
- purchasing water rights at the expense of healthy regional communities and moving agriculture north
- planning for the expansion of regional growth centres and planning for the water needs of large cities in 2050
- planning for the expansion of major cities and urban centres
- planning for a sustainable agricultural sector.

Some Caveats

We are acutely aware that only a small subset of people with expertise in land use policy and planning participated in our June 2015 meeting; an even smaller number contributed a chapter to this book. Consequently, we acknowledge that there will be perspectives that are not represented, either in part or in full, in this volume. In many respects, this is positive, for it means that there is more to be said and written about land use policy and planning, and how information and tools can be better generated, managed and used to improve policy and decision-making. If this book stimulates additional dialogue that fosters support for better informed land use policy and planning, we believe that the exercise will have been a valuable one.

References

Steffen W., Burbidge A., Hughes L., Kitching R., Lindenmayer D., Musgrave W., … Werner P. (2009). *Australia's biodiversity and climate change: A strategic assessment of the vulnerability of Australia's biodiversity to climate change.* A report to the Natural Resource Management Ministerial Council commissioned by the Australian Government, CSIRO Publishing. Retrieved from www.climatechange.gov.au/sites/climatechange/files/documents/042013/biodiversity-vulnerability-assessment.pdf

Thackway, R. & Lesslie, R. (2008). Describing and mapping human-induced vegetation change in the Australian landscape. *Environmental Management 42*, 572–90. doi.org/10.1007/s00267-008-9131-5

2

A Retrospective: The Influence of Rob Lesslie—Landscape Scientist, Geographer and Natural Resources Land Use Planner

Richard Thackway

Key Points

- Effective leadership in the areas of land use policy and planning involves understanding the values and ideals of others. Respected leaders engage the key players, listen well and practise diplomacy.

- The process of developing, implementing and evaluating land use policies and plans involves leaders collecting, collating and curating large amounts of relevant spatial and temporal data and information. Key leaders of land use planning programs demonstrate the value of primary data: collect once, use many times.

- Sound leadership knows how to use conceptual models that represent the real world, assisting decision-makers and other stakeholder groups to visualise complex eco-social systems, investigate and manipulate data and information, and discover and inform collective decision-making (e.g. what is now, what could be and what can be).

- Developing realistic scenarios of alternative or different social, economic and environmental outcomes, both in space and time, requires leaders with humility and strength. Influential leaders participate in the planning process and are true to their core principles.

- Promoting long-term sustainable management and conservation of Australia's national landscapes and ecosystems involves influencing future leaders now; influential leaders are generous with their resources, including philanthropy.

These key lessons are discussed via a retrospective of the influence of Dr Robert George Lesslie—landscape scientist, geographer and natural resources land use planner. Among Rob's many interests and passions was his desire to identify and preserve the wild places and undisturbed natural areas of Australia. I learned about this passion over a 20-year period, as we periodically walked and talked. Rob's contributions to the fields of land use policy and planning were impressive. This retrospective presents representative examples of his life's work.

Robert George Lesslie was born in Sydney on 25 February 1957; he lived in Adelaide and Canberra, which is where he died on 28 March 2014. Emeritus Professor Henry Nix observed:

> Future generations will come to value his research contributions to key questions of conservation, land management and sustainability. Rob was just reaching the pinnacle of his creative contributions and Australia and the world is the poorer for his loss (Henry Nix, personal communication).

Rob's career spanned more than 30 years; he worked in natural resources evaluation and management in government, education and the private sector. Throughout his career, he published numerous book chapters, journal articles and technical reports. He was a sought-after speaker at conferences and workshops. However, Rob's influence went well beyond this, as he was also significantly involved in the formulation and implementation of pressing natural resource management policies.

I first met Rob in the late 1980s when I was appointed an advisor on the Australian Government Scientific Review Committee, convened to evaluate the National Wilderness Mapping Project. From that time on, our paths crossed regularly on various projects, including vegetation condition, land management and rangeland management, ecosystem services and the potential to combine geographic information systems (GIS) and Bayesian belief networks for environmental assessment and classification based on the vegetation assets, states and transitions (VAST) framework.

Rob's tertiary studies and research focused on how humans interact with the environment and the effects of use and management of (Australia's) landscapes. He gained a Bachelor of Arts from Macquarie University in 1978, majoring in geography and economics, and a Master of Environmental Studies from the University of Adelaide in 1982. In 1997, he was awarded a PhD from The Australian National University for his thesis entitled 'A spatial analysis of human interference in terrestrial environments at landscape scales'.

Rob profoundly influenced both the assessment and mapping of land use and ecological condition and the use of this information in the development of land use policy, natural resource management and conservation in Australia. I am grateful to have learned the following key lessons from him.

Practise Diplomacy

Rob's ability to develop genuine relationships in the course of his professional associations was outstanding. Easygoing and easy to get along with, he was widely valued as a friend. Staff and colleagues commented that they felt he took a real interest in their work and that he listened carefully to them and genuinely considered the matters they raised. These characteristics travelled with him—in intergovernmental meetings, sectoral meetings with industry, his interactions with government officials and scientific debates—and he demonstrated a respect for others. As a result, he was welcomed and valued in these spheres.

Rob's character was clearly expressed through his involvement in and coordination of the Australian Collaborative Land Use and Management Program (ACLUMP, 2010; National Land and Water Resource Audit [NLWRA], 2008). A partnership between state, territory and federal government land management agencies, the program provides consistent land use information across the whole continent. Partners contribute to broader national understandings on a wide range of natural resource management (NRM) subjects, including the management and protection of native vegetation, water and soil conservation and production, and the control of weeds and pest animals. This involves working together to establish national guidelines for nationally consistent land use mapping coverage for Australia at 'continental' and 'catchment' scales; developing a national information system for land management practices—the 'how' of land use; developing agreed national technical standards, including the Australian Land Use and Management Classification; developing a national land use data directory and the maintenance of land use datasets on federal and state government data repositories; and developing regional and national systems for reporting land use and land management practices (NLWRA, 2008).

Before Rob began coordinating ACLUMP, each member agency had disparate ways of measuring and collecting land use data; there was no means by which an overall understanding of Australian land use practices could be assessed and reported, or an Australia-wide land use map produced.

Gaining the support of Australia's various state and federal land management agencies required considerable diplomacy; it called for someone committed to collaboration and with a strong sense of fairness— someone of integrity, generosity, commitment and humility: Rob. Rob's ability to gain support and achieve consensus among various parties grew out of his respect for the different roles and responsibilities of state and federal agencies within a national context. He was able to identify common strategies and priorities that facilitated joint investment in a collaborative land use mapping program. The cohesion of the partnership was due, in no small measure, to Rob's considerable intellectual and people-based aptitude.

One of the few surviving national coordinating committees, ACLUMP is recognised as an exemplar in cross-jurisdictional natural resource coordination. Rob supported a culture of regular meetings in which land use policy and planning issues could be openly discussed. Such issues included how industry sectors might better contribute to national ecological sustainable development, the need for common land use classification frameworks to enable the analysis of information to meet immediate and emerging policy issues, and the need to investigate new technologies so that changes in land use and the environmental effects could be more rapidly monitored.

Rob's insights and writings continue in the work and research of other individuals across Australia. His commitment to land use policy and planning is seen in others who are now fostering and promoting the collection and dissemination of nationally consistent land use and land management practices data and information. It is also reflected by those who recognise the benefits of working with key decision-makers to use information to influence land use policy and planning decisions.

Demonstrate the Value of Primary Data: Collect Once, Use Many Times

In relation to the Australian National Wilderness Inventory (ANWI), a regional-scale digital mapping and assessment project that began in 1986, I recall asking Rob about the data collection standards he was using, and whether the states and territories were supplying primary digital data, or if he was collecting it himself. His answers were somewhat surprising.

States and territories that supported the philosophy of the wilderness mapping project provided in-kind support to digitally capture the relevant datasets from printed documents. When digitised spatial data, such as road and track networks, forest management practices and watering points in rangelands, were not supplied, Rob developed them himself by collecting and analysing large amounts of primary digital data. Systematic and well organised, he then provided these data to the relevant jurisdictions.

It is worth noting that there was considerable debate at the time about the appropriateness of using the term 'wilderness', given past and current Indigenous occupation and management of Australia. It was also a time when most land management agencies were just beginning to embrace computers to analyse and store NRM data, and commencing the process of digitally converting paper maps and reports into computer-readable data formats.

It is worth considering some of the technical constraints that Rob faced, and his foresight in building an enduring data infrastructure that was designed to inform and influence future land use policies and planning concerning wilderness, its type, extent and modification states. Compared to what we have today, his operating environment was extremely clunky and primitive, to say the least. For example, GIS were command-driven; Rob's early adoption of GIS to collect, collate and store spatial data, and to assess the effects of land use on wilderness quality at a regional scale across Australia, was exemplary. His use of regional and national databases was essential, as these enabled him to operate consistently across jurisdictions at a regional scale. His applications of these awkward and cumbersome tools were recognised as exemplary by academic colleagues and those in government. Digitising was manual, slow and tedious. Metadata for spatial datasets was neither developed nor accepted as good practice; standards were neither defined nor adopted nationally—these took another 10 years. Yet, Rob meticulously documented these datasets. Printers were evolving from hexadecimal and symbol-based printing, but shades of grey were what was required. Rob was a pioneer in the use of GIS to present the big picture.

As the wilderness project progressed, so too did the power and sophistication of GIS and databases. Given the opportunity to migrate primary data from one platform to the next, Rob was quick to take advantage of developments in flexibility and interactivity (Lesslie, 2016; Lesslie & Maslen, 1995). These developments, which allowed for improved analysis and visualisation of wilderness over large areas, enhanced the ability to understand regional issues and develop regional solutions.

The Multi-Criteria Analysis Shell for Spatial Decision Support (MCAS-S) tool represents the culmination of Rob's support for the efficient use of primary data across multiple projects over time (Lesslie, 2013; Lesslie et al., 2008). The MCAS-S has been widely used by government agencies, NRM groups, land use policy and planning researchers and agricultural scientists. The tool provides ready access to large quantities of environmental, social and economic information, and has straightforward analytical features. The MCAS-S helps decision-makers to understand and visualise the spatial and temporal patterns of their world, analyse biophysical and cultural interactions, quickly investigate patterns and processes and provide a full record of the decisions made.

Use Conceptual Models to Represent Complex Eco-Social Systems

Rob had a well-developed ability to synthesise disparate sources of data to produce novel insights into complex situations. The landscape we see today is the product of millennia of evolution and change, usually incorporating some level of human use and management. The human use of landscapes varies from intensive and localised to extensive and widespread. The frequency, duration and magnitude of the effects of human use are different in various ecosystems. Land managers employ a number of management practices to maintain or change the ecological function of native vegetation at site or landscape levels. This number increases when historic practices are added to the list of contemporary practices (Thackway & Specht, 2015). Rob was cognisant of these complexities and often constructed conceptual models to convey concise and simple messages to decision-makers.

Rob's geographical training influenced his approach to communication. He concentrated on using spatial models and maps to provide clear visual information, and used real-world examples to illustrate the effects of land use on landscape patterns and processes. His research into eco-social systems—including pastoralism in the arid rangelands, managed production forests and intensively managed cropping and pastures—generated insights into the responses of ecosystems to different management regimes and threats. His breadth and depth of experience was highly valued by managers of policy and programs in considering the design and implementation of public programs to achieve good public policy outcomes.

As a result of his doctoral research, Rob was aware that those places that remained relatively remote and natural were becoming increasingly rare and more valuable as modern technological society extended its reach and impact across Australia (Lesslie, 2016). Wilderness is context dependent—it is relative. Defining wilderness quality as a continuum of remote and natural conditions, and using assessable units and a common scale, provides decision-makers with a solid conceptual foundation to approach the issue of identifying wilderness resources. It also provides a coherent evidence base for discussion and debate regarding wilderness more broadly—from concerns about its cultural context, through to measures for wilderness protection and management (Lesslie, 2016). This focus on remote and natural conditions enables decision-makers to quantify thresholds of interest and to visualise these as effects of modern technological societies in landscapes (i.e. patterns of land use and management, and settlement and access). The conceptual model for wilderness value is shown in Figure 2.1.

The capacity to represent this conceptual model as mapped spatial layers provides decision-makers with unprecedented opportunities to identify and track spatial and temporal changes in the type, extent and modification of wilderness (see Figure 2.2). Further, newly available data streams, including remotely sensed land cover data, digital topographic and terrain mapping and land use and dynamic vegetation, can be used to enhance the mapping of wilderness quality.

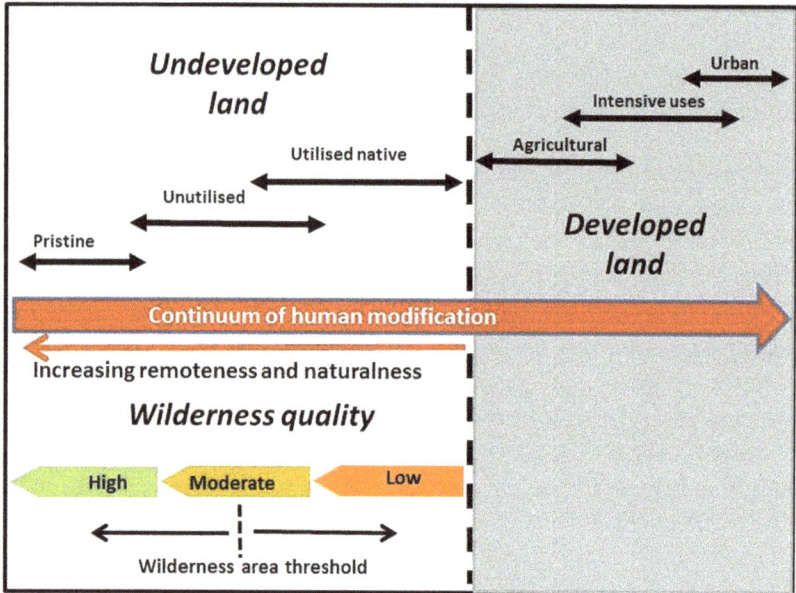

Figure 2.1: The wilderness continuum concept.
Source: Adapted from Lesslie & Taylor (1985).

A similar conceptual model based on degrees of human modification of plant communities is the VAST framework (Lesslie, 2016). Maps produced representing this model inform decision-makers about the extent of native vegetation and changes in condition at regional and national levels (Thackway & Lesslie, 2008). By combining maps of vegetation type (Department of the Environment and Water Resources, 2007) with the condition of vegetation (e.g. Figure 2.3), decision-makers gain powerful insights into setting national and regional priorities for the restoration and rehabilitation of highly modified landscapes (e.g. Yapp & Thackway, 2015, Figure 7).

Figure 2.2 shows the distribution of wilderness quality across Australia based on the results of the ANWI (the survey was incomplete in far south-western Australia) (Lesslie & Maslen, 1995). The threshold at which 'wilderness' is recognised changes according to environmental context and over time. Figure 2.3 shows similar areas to those delineated as wilderness (Figure 2.2) using a set of area selection criteria and additional assessments to validate and revise the ANWI results (Department of Sustainability, Environment, Water, Population and Communities, 2008).

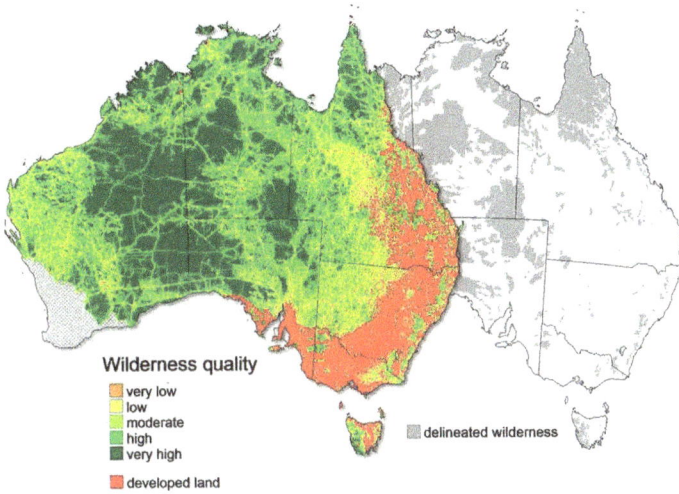

Wilderness quality
- very low
- low
- moderate
- high
- very high

- developed land

delineated wilderness

Figure 2.2: Wilderness quality and delineated wilderness in Australia, c. 1990.

Source: Lesslie (2016).

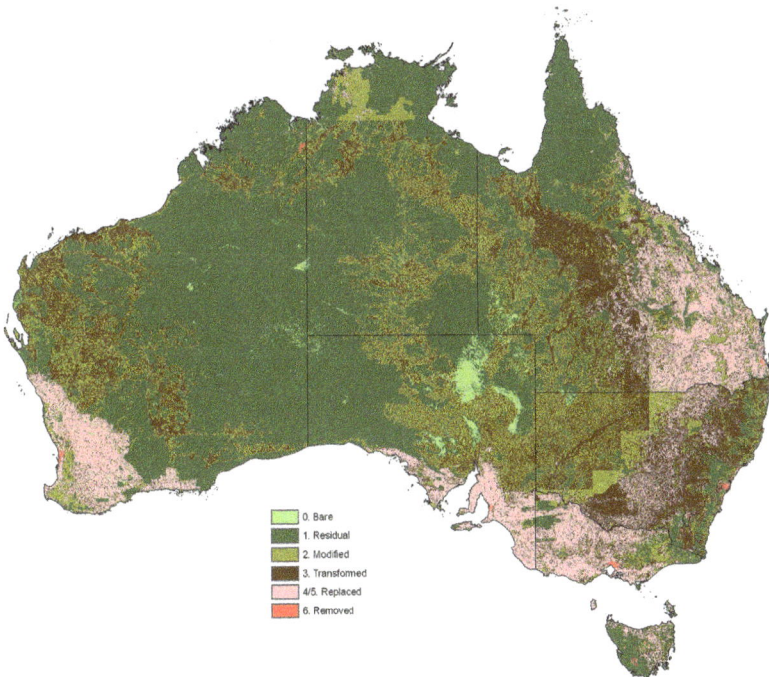

- 0. Bare
- 1. Residual
- 2. Modified
- 3. Transformed
- 4/5. Replaced
- 6. Removed

Figure 2.3: Vegetation assets, states and transitions (VAST) dataset for Australia (version 2.0).

Source: Lesslie, Thackway & Smith (2010).

Be True to Your Core Principles

Rob felt strongly that, in addition to the spatial, temporal scales in land use planning were vital. There is an imperative to move beyond the apparent current short-term political cycle that drives inappropriate resource management decisions to a longer-term intergenerational approach informed by the best available information (data) at a national level.

In 2007, Rob was a founding member of the ongoing Wilderness Society WildCountry Science Council (Mackey et al., 2007; Wilderness Society n.d.-a). This group has four pillars that represent recurring themes in Rob's life's work: ecological science, Indigenous conservation, climate change and economic development. One of the group's core values is to maintain and protect natural processes that sustain human wellbeing, biota and landscapes across the continent. Echoing the worldview that Rob espoused in his private life and public work, the Wilderness Society's (n.d.-b) vision is to transform Australia into a society that protects, respects and connects with the natural world that sustains us. The WildCountry Science Council's activities in promoting a whole-of-landscape approach, and focusing campaigns and policies on the importance of ecological connectivity at both regional and continental scales, also resonate with Rob's work on wilderness quality as a continuum to provide evidence for land use policy and planning.

Regrettably, Rob's involvement in the WildCountry Science Council was deemed incompatible with the roles and responsibilities of a public servant in the Australian Government. He resigned his membership of the council but continued to observe its work from a distance. It is worth noting that the formative work of this council provided the fundamental underpinning of the landscape connectivity movement that is now mainstream across regional Australia (Fitzsimons et al., 2013).

Influence Future Leaders of Land Use and Planning Through Philanthropy

A focus of Rob's last year of life was to work with his family in establishing a lasting philanthropic legacy to support future research and other land management activities at the Fenner School of Environment and Society at The Australian National University. Rob's aim was the promotion of long-term sustainable management and conservation of Australia's national landscapes and ecosystems. The Lesslie Endowment (n.d.) will provide support for research grants, scholarships, fellowships, prizes or public seminars and workshops.

Acknowledgements

Lynne Alexander, Jake Gillen and Graham Yapp provided helpful advice on the structure and content of an earlier draft.

References

ACLUMP (Australian Collaborative Land Use and Management Program). (2010). *Land use and land management information for Australia: Workplan of the Australian Collaborative Land Use and Management Program.* Canberra, ACT: Australian Bureau of Agricultural and Resource Economics and Sciences.

Department of the Environment and Water Resources. (2007). *Australia's Native Vegetation: A summary of Australia's Major Vegetation Groups, 2007.* Australian Government, Canberra, ACT.

Department of Sustainability, Environment, Water, Population and Communities. (2008). *Remote and natural lands delineation.* Retrieved from www.environment.gov.au/heritage/publications/anlr/maps-delin.html

Fitzsimons, J., Pulsford, I. & Wescott, G. (Eds.). (2013). *Linking Australia's landscapes: Lessons and opportunities for large-scale conservation networks.* Melbourne, VIC: CSIRO Publishing.

Lesslie Endowment. (n.d.). Retrieved from www.anu.edu.au/giving/support-us/lesslie-endowment

Lesslie, R. (1997). *A spatial analysis of human interference in terrestrial environments at landscape scales* (Unpublished doctoral dissertation). The Australian National University, Canberra, ACT.

Lesslie, R. (2013). Mapping our priorities—innovation in spatial decision support. In P. Figgis, J. Fitzsimons & J. Irving (Eds.), *Innovation for 21st century conservation* (pp. 156–63). Sydney, NSW: Australian Committee for the International Union of Conservation of Nature.

Lesslie, R. (2016). The wilderness continuum concept and its application in Australia: Lessons for modern conservation. In S. Carver & S. Fritz (Eds.), *Mapping wilderness: Concepts, techniques and applications* (pp. 22–40). New York, NY: Springer. doi.org/10.1007/978-94-017-7399-7_2

Lesslie, R. G., Hill, M. J., Hill, P., Cresswell, H. P. & Dawson, S. (2008). The application of a simple spatial multi-criteria analysis shell for natural resource management decision making. In C. Pettit, W. Cartwright, I. Bishop, K. Lowell, D. Pullar & D. Duncan (Eds.), *Landscape analysis and visualisation: Spatial models for natural resource management and planning* (pp. 73–96). Berlin: Springer. doi.org/10.1007/978-3-540-69168-6_5

Lesslie, R. G. & Maslen, M. (1995). *National wilderness inventory, Australia: Handbook of procedures, content and usage* (2nd ed.). Canberra, ACT: Australian Heritage Commission.

Lesslie, R. G. & Taylor, S. G. (1985). The wilderness continuum concept and its implications for Australian wilderness preservation policy. *Biological Conservation, 32,* 309–33.

Lesslie, R., Thackway, R. & Smith, J. (2010). *A national-level vegetation assets, states and transitions (VAST) dataset for Australia* (version 2.0). Canberra, ACT: Bureau of Rural Sciences. Retrieved from data.daff.gov.au/data/warehouse/pe_brs90000004193/VASTv2Data_20100320_ap14.pdf

Mackey, B. G., Soulé, M. E., Nix, H. A., Recher, H. F., Lesslie, R. G., Williams, J. E., … Possingham, H. P. (2007). Applying landscape-ecological principles to regional conservation: The WildCountry Project in Australia. In J. Wu & R. J. Hobbs (Eds.), *Key topics and perspectives in landscape ecology* (pp. 192–213). Cambridge, UK: Cambridge University Press.

NLWRA (National Land and Water Resources Audit). (2008). *Land use— status of information for reporting against indicators under the national resource management monitoring and evaluation framework.* Canberra, ACT: National Land and Water Resources Audit.

Thackway, R. & Lesslie, R. (2008). Describing and mapping human-induced vegetation change in the Australian landscape. *Environmental Management, 42,* 572–90. doi.org/10.1007/s00267-008-9131-5

Thackway, R. & Specht, A. (2015). Synthesising the effects of land use on natural and managed landscapes. *Science of the Total Environment, 526,* 136–52. doi.org/10.1016/j.scitotenv.2015.04.070

Wilderness Society. (n.d.-a). *WildCountry related scientific publications.* Retrieved from www.wilderness.org.au/wildcountry-related-scientific-publications#sthash.H0Nv1SXY.dpuf

Wilderness Society. (n.d.-b). *Our purpose.* Retrieved from www.wilderness. org.au/our-vision

Yapp, G. A. & Thackway, R. (2015). Responding to change—criteria and indicators for managing the transformation of vegetated landscapes to maintain or restore ecosystem diversity. In J. A. Blanco (Ed.), *Biodiversity in ecosystems—linking structure and function.* Retrieved from www.intechopen.com/books/biodiversity-in-ecosystems-linking-structure-and-function/responding-to-change-criteria-and-indicators-for-managing-the-transformation-of-vegetated-landscapes

3

Reprint: Land Use and Management—The Australian Context[1]

Rob Lesslie and Jodie Mewett

Key Points

- Over the last 100 years there is an overall trend of land use intensification in Australia, although this varies regionally and is set against a slow decline in the proportion of Australia's land area used for agriculture.

- Australia's food supply system will continue to contribute to food security domestically and globally.

- Under the Australian Constitution, the state governments have prime responsibility for land administration and public land management. The Australian Government has a limited land ownership and management role. Its primary role is to promote more efficient land management and land allocation.

- Important themes of governance at the state level include urban and rural land zoning, forestry plantations, mining development and environmental regulations around native vegetation management.

---

1   R. Lesslie & J. Mewett. (2013). *Land use and management: The Australian context* (Australian Bureau of Agricultural and Resource Economics and Sciences Research Report 13.1). Canberra, ACT. Retrieved from agriculture.gov.au/abares/publications.

- Improved data collection and analytical systems are needed for tracking 'hotspots' of land use change, including land use intensification on the fringes of cities and urban areas, and loss of productive agricultural land, productive land resources and biodiversity.
- Projected increased population will create further pressure for land use intensification for residential, commercial and production purposes.
- Australian and state governments have implemented a mix of programs and regulations to enhance land management and land use.

Introduction

Australia has unique land, water, vegetation and biodiversity resources. Australia's 7.7 million square kilometres support a wide range of agricultural and forestry industries. Production from natural resources earns over $38 billion a year in exports from agriculture, fisheries and forestry. Competitive pressures drive the need for improved productivity, which includes increased diversification and intensification. These trends are occurring against a background of increased climate variability.

The way in which land is used has a profound effect on Australia's social and ecological systems. There is a strong link between changes in land use and environmental, economic and social conditions. Information on land use and management is fundamental to understanding landscapes, agricultural production and the management of natural resources.

Land use choices have a major effect on our food production, natural environment and communities. Land use change and land management are central to current debate in Australia around food security, water, climate change adaptation, population and urban expansion. Informed land use and land management choices are critical to developing effective responses to natural resource management priorities, such as biodiversity protection, sustainable and productive agriculture, water quality and quantity, salinity and food security.

Key Land Statistics

The pattern of land use in Australia is shown in Figure 3.1. The dominant land use is livestock grazing. This occurs mostly on native vegetation and makes up 56 per cent (or 4.3 million square kilometres) of Australia (see Table 3.1). Other agricultural uses, including broadacre cropping (almost 270,000 square kilometres or 3.5 per cent) and horticulture

(5,000 square kilometres or less than 0.1 per cent), occupy a much smaller proportion of land area. The total area of land under primary production in Australia (livestock grazing, dryland and irrigated agriculture) is nearly 4.6 million square kilometres or 59 per cent.

Approximately 569,240 square kilometres (or 7 per cent) of Australia is set aside for nature conservation. Other protected areas, including those for use by Indigenous Australians, cover more than 1 million square kilometres (or 13 per cent) of Australia. Forestry tends to be confined to regions with higher rainfall and covers nearly 2 per cent of the continent. Intensive uses (mostly urban) occupy about 17,000 square kilometres (or 0.2 per cent) of Australia.

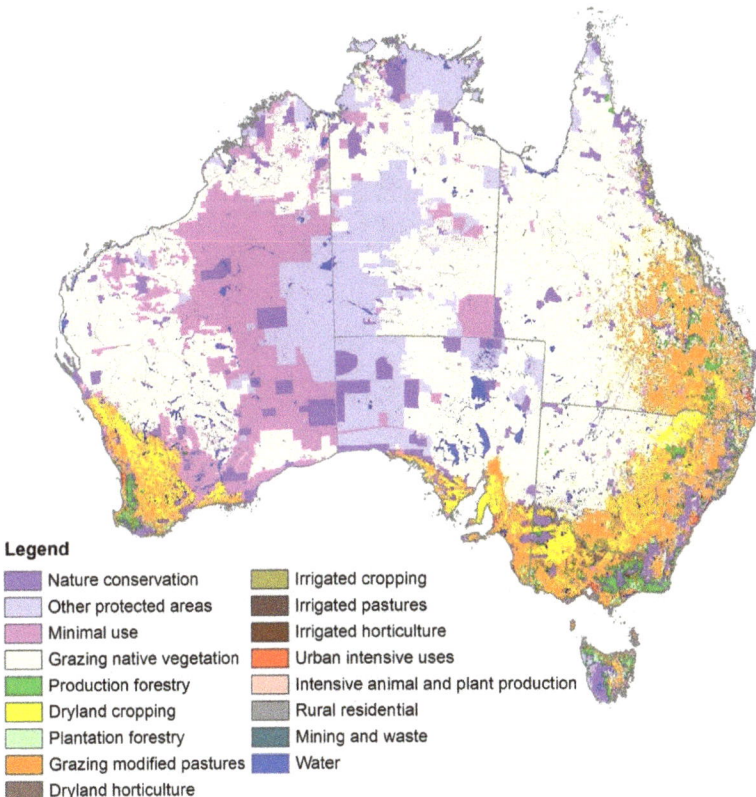

Legend

Nature conservation		Irrigated cropping	
Other protected areas		Irrigated pastures	
Minimal use		Irrigated horticulture	
Grazing native vegetation		Urban intensive uses	
Production forestry		Intensive animal and plant production	
Dryland cropping		Rural residential	
Plantation forestry		Mining and waste	
Grazing modified pastures		Water	
Dryland horticulture			

Figure 3.1: National land use of Australia, 2005–06.[2]

Source: Bureau of Rural Sciences (2006).

2   Note: Since the original publication of this paper, a more recent land use product has been released for 2010–11. To view the updated map of land use of Australia, 2010–11, see www. agriculture.gov.au/abares/aclump/land-use/land-use-mapping.

Table 3.1: Land use in Australia, 2005–06.

Land use	Area km²	Proportion of Australia %
Grazing native vegetation	3,558,785	46.30
Minimal use	1,242,715	16.17
Other protected areas, including Indigenous uses	1,015,359	13.21
Grazing modified pastures	720,182	9.37
Nature conservation	569,240	7.41
Dryland cropping	255,524	3.32
Water	125,618	1.63
Production forestry	114,314	1.49
Plantation forestry	23,929	0.31
Intensive uses (mainly urban)	16,822	0.22
Irrigated cropping	12,863	0.17
Irrigated pastures	10,011	0.13
Rural residential	9,491	0.12
Irrigated horticulture	3,954	0.05
Intensive animal and plant production	3,329	0.04
No data	2,243	0.03
Waste and mining	1,676	0.02
Dryland horticulture	1,092	0.01
Total	7,687,147	100.00

Source: Australian Bureau of Agricultural and Resource Economics and Sciences— Bureau of Rural Sciences (2010).

History

The Aboriginal occupation of Australia has been associated with the systematic burning of vegetation to increase the availability of plant and animal foods and reduce fuel for wildfires (Gammage, 2011). Since European settlement about two centuries ago, Australia's landscapes have changed significantly. European settlement began with early pastoralism, cropping and prospecting, and has led to today's major agricultural, forest and mining industries, reserve landscapes and urban communities. Land use change over this period was driven by relatively unrestricted access to land, technological change and growth in productivity and population.

More recently, land has been increasingly managed for multiple objectives, including food, fibre, minerals, energy, landscape amenity, water, carbon and biodiversity. A well-managed landscape provides high-quality and essential ecosystem services to farmers and the Australian community. Australian governments have implemented a mix of programs and regulations to enhance land management and use. The Australian Government has also invested in programs to improve land management practices, such as the Natural Heritage Trust, the National Action Plan for Salinity and Water Quality and Caring for our Country.

Current Situation

Policy–Legislation Framework

Significant areas of Australia are used for livestock grazing on native pastures. This may be freehold or public land leased by private landholders. Agricultural uses—including cropping and livestock grazing on improved pastures, urban uses, other intensive uses and Indigenous uses—are mainly located on privately owned (freehold) land. Most land allocated to forestry (native forests) and nature conservation is publicly owned and managed.

Australia has six states and two mainland territories. Under the Australian Constitution, the state governments have primary responsibility for land administration and public land management. The Australian Government has a limited land ownership and management role. Its primary role is to promote more efficient land management and land allocation. An example is the National Forest Policy process, which has established a nationally agreed basis for determining forest assessment and resource allocation principles, resource inventory and national reporting. There are similar national coordination processes for nature conservation; for example, through the National Reserve System. Important themes of governance at the state level include urban and rural land zoning, forestry plantations, mining development and environmental regulations around native vegetation management.

There are three key challenges:

1. Land use governance and tenure arrangements need to accommodate multi-objective land use and land management options. Modern approaches to nature conservation include joint management and partnership arrangements that can simultaneously provide for multiple uses; these can include biodiversity protection, recreation uses, Indigenous cultural uses, mineral exploration, mining and grazing production.

2. Land use and land management incentives for non-market ecosystem goods and services may play an important role; these may be applied where these goods and services are not adequately reflected in market systems.

3. Sufficient data and information are required to support informed land use and management planning. This includes decision support capability to enable informed and transparent consideration of options and trade-offs.

Agricultural Land Management

In 2007–08, the Australian Bureau of Statistics (ABS) surveyed farmers about their land management practices. Table 3.2 shows characteristics of Australian farms broken into four major industries: broadacre cropping, horticulture, dairying and grazing (beef cattle, sheep meat).

Innovation in Australian agriculture has generally driven enough growth in productivity to offset consistently declining agricultural terms of trade. Innovation, through improved land management practices, has also increased agriculture's ability to lessen threats to soil, water resources and biodiversity. Land management practices can bring about the changes needed for sustainable use of Australian landscapes. For example, conservation tillage helps to improve soil carbon, reduce soil erosion and nutrient loss, and increase cost savings and other production benefits. Of agricultural businesses preparing land for crops or pastures, 40,000 (53 per cent) reported using no tillage over a total of more than 170,000 square kilometres in 2007–08. Figure 3.2 shows how tillage management practices differ across the country.

Table 3.2: Summary of characteristics and management of Australian farm businesses (farmers).

Farm business characteristics and management	Broadacre cropping	Horticulture	Dairying	Grazing (beef cattle, sheep meat)
Farm business characteristics				
Average age	54	53	53	55
Average years managing holding	24	19	22	23
Farm management plans (formal and informal)				
Production management	56,891 (81%)	13,934 (59%)	8,270 (83%)	75,579 (82%)
Business financial	51,371 (73%)	15,490 (65%)	7,796 (79%)	64,734 (70%)
Extreme circumstances	47,800 (68%)	12,994 (55%)	7,502 (76%)	64,123 (69%)
Natural resource management	42,215 (60%)	11,887 (50%)	6,557 (66%)	56,537 (61%)
Succession	39,593 (57%)	10,705 (45%)	6,055 (61%)	49,996 (54%)

Note: Numbers in brackets represent the proportion of farmers adopting farm plans by enterprise type.
Source: Australian Bureau of Statistics (ABS, 2008).

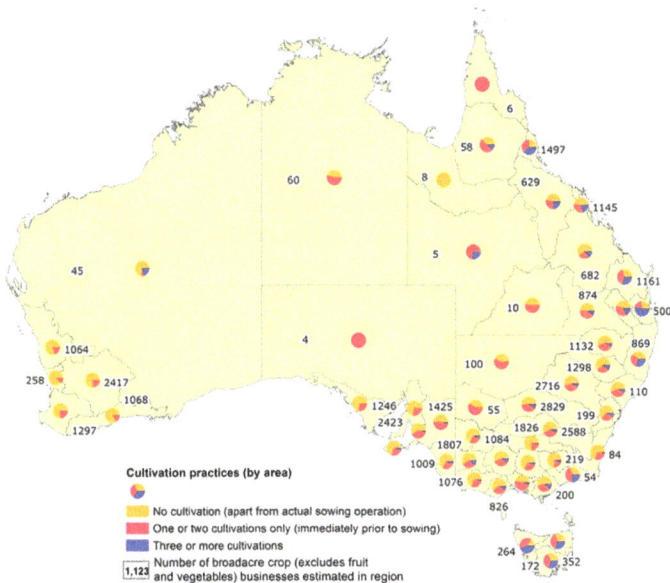

Figure 3.2: Cultivation practices by natural resource management region.
Note: Refers to cultivation practices used to prepare crops and pastures as a percentage of the area prepared for crops and pastures, by natural resource management region.
Source: ABS (2009).

Natural Resources

Water

Long-term average annual rainfall varies across Australia from less than 300 millimetres per year in most of Central Australia to more than 3,000 millimetres per year in parts of Far North Queensland (see Figure 3.3).

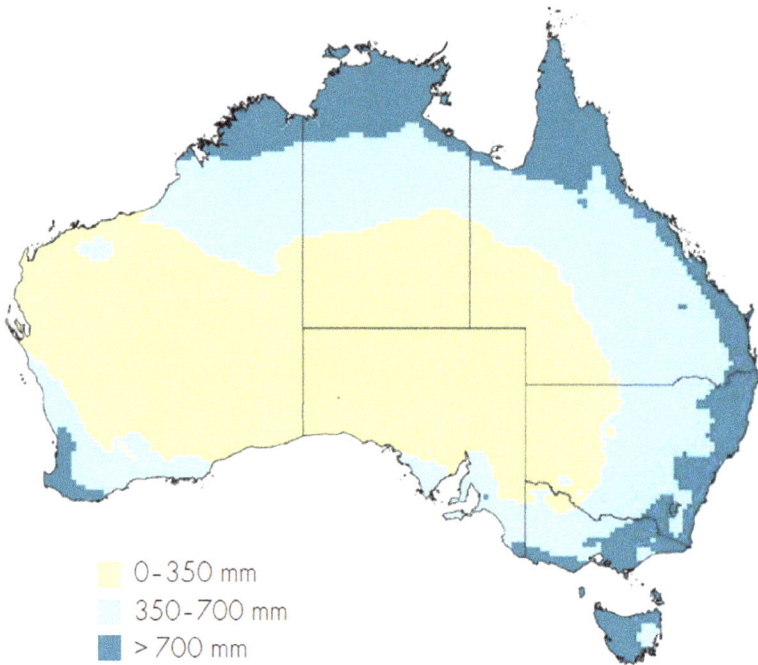

Figure 3.3: Mean annual rainfall of Australia.
Source: Department of Agriculture, Fisheries and Forestry (DAFF, 2008).

Of all inhabited continents, Australia has the lowest proportion of rainfall going into its rivers and aquifers—11 per cent compared with the world average of 65 per cent. About 65 per cent of run-off occurs in far north Australia and coastal Queensland (see Figure 3.4). By contrast, only 6.8 per cent of Australia's run-off occurs in the Murray–Darling Basin, although more than 50 per cent of Australia's water use occurs there. The seasonal distribution of rainfall also varies widely—run-off in northern Australia occurs predominantly in the monsoonal wet season, while run-off in the Murray–Darling Basin is spread throughout the year.

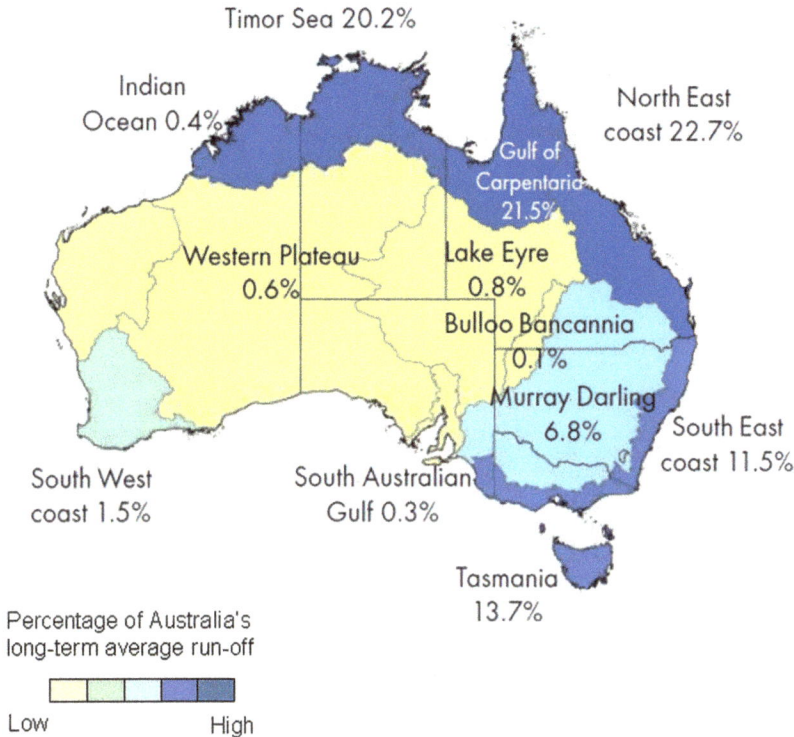

Figure 3.4: Long-term average run-off from Australia's drainage divisions.
Source: DAFF (2012).

Irrigation is a well-established and important feature of the agricultural landscape, especially in the Murray–Darling Basin. In 2007–08, 90 per cent (6,285 gigalitres) of the water used by agricultural industries was used to irrigate crops and pastures, while 10 per cent (704 gigalitres) was used for other agricultural purposes. Water use for irrigation, industry and urban needs has placed pressure on water-dependent ecosystems. The challenge is using water for production purposes while maintaining water quality and conserving the natural environment. A national program of water reform is being implemented to achieve this balance.

Soils

The agricultural landscape in Australia includes a wide range of soil types, ranging from old, deeply weathered and infertile soils to younger and more fertile soils (see Figure 3.5) (McKenzie et al., 2004).

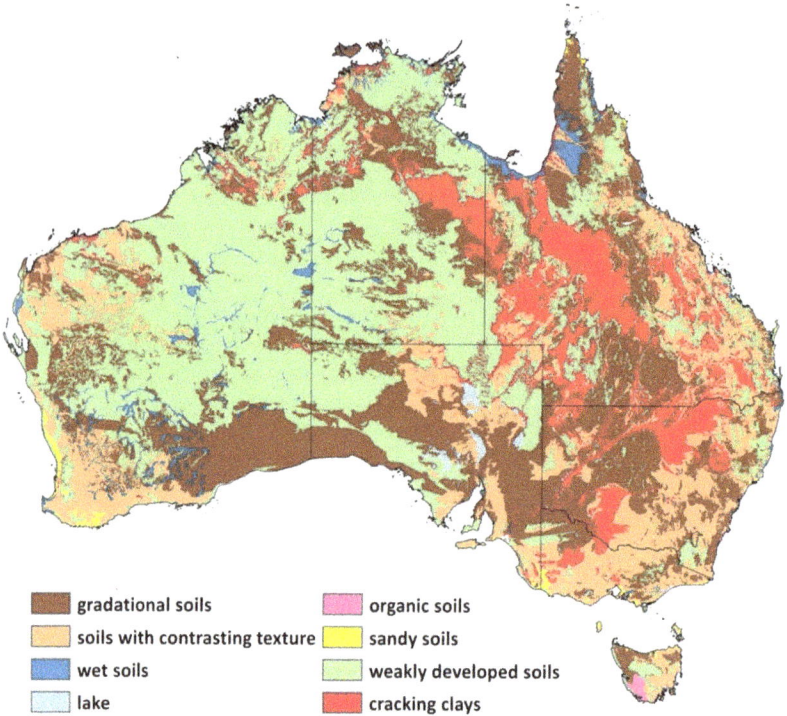

Figure 3.5: Australian soil types.
Source: DAFF (2012).

Important soil management issues include erosion (wind and water), salinisation (dryland and irrigation), acidification and compaction. Soils are managed by maintaining ground cover and windbreaks, avoiding steep slopes, applying fertilisers (mainly phosphorus and nitrogen) and using lime and gypsum to manage soil condition and pH.

Vegetation

Australia's native vegetation estate comprises shrublands and heathlands (37 per cent), native grasslands and minimally modified pastures (33 per cent) and native forests and woodlands (19 per cent). The remaining 10 per cent of the continent comprises non-native vegetation, such as annual crops and modified pastures (9 per cent) and plantations (0.2 per cent) (see Figure 3.6).

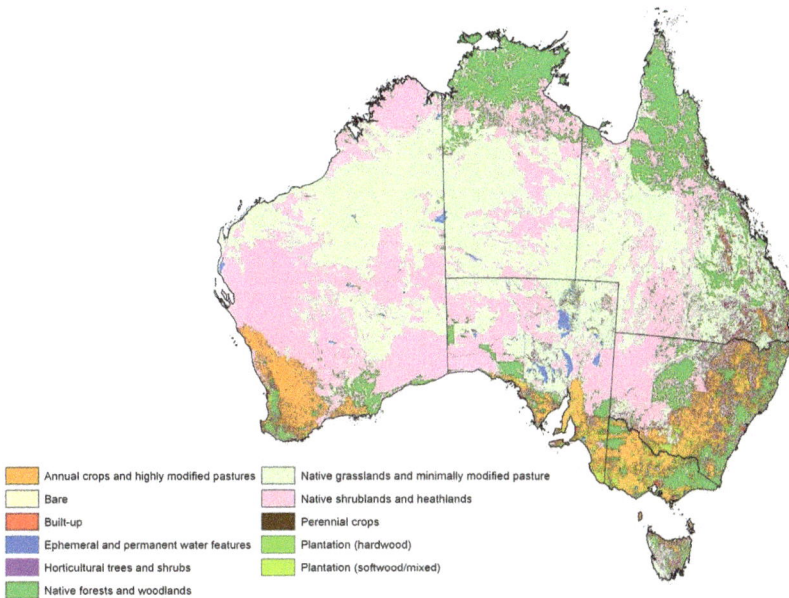

Figure 3.6: Australian vegetation.
Source: DAFF (2012).

In areas with higher rainfall and more fertile soils, native vegetation has been extensively cleared and replaced with intensive agriculture. In these landscapes, remnants of native vegetation and non-native vegetation exist as a mosaic of vegetation types. In areas with lower rainfall, much of the native vegetation remains and supports pastoral industries. Providing water (i.e. bores) has enabled extensive development of these rangelands.

The clearing of native vegetation has declined since the 1990s, as states and territories introduced regulatory controls. Vegetation management can profoundly affect landscape condition, soil health and the supply of ecosystem goods and services, such as food, fibre and water production. The importance of managing vegetation for ecosystem goods and services is reflected in on-the-ground investments being made through conservation programs, land clearing controls and environmental management systems.

Environment Protection

Government Programs

The Australian Government's central piece of environmental legislation is the *Environment Protection and Biodiversity Conservation Act 1999* (EPBC). It provides a legal framework for protecting and managing flora, fauna, ecological communities and heritage places that are defined as matters of national environmental significance. The objectives of the EPBC are to:

- provide for the protection of the environment, especially matters of national environmental significance
- conserve Australian biodiversity
- provide a streamlined national environmental assessment and approvals process
- enhance the protection and management of important natural and cultural places
- control the international movement of plants and animals (wildlife), wildlife specimens and products made or derived from wildlife
- promote ecologically sustainable development through the conservation and ecologically sustainable use of natural resources.

Environment protection objectives are also pursued by the Australian Government under the Caring for our Country initiative, which supports the environmental management of natural resources. The Caring for our Country initiative addresses six national priority issues:

1. expanding Australia's National Reserve System
2. addressing threats to biodiversity and natural icons
3. managing and protecting coastal environments and critical aquatic habitats, including the Great Barrier Reef
4. encouraging farmers to adopt sustainable farm practices
5. promoting sustainable natural resources use and environmental protection in northern and remote Australia
6. supporting and increasing community skills, knowledge and engagement.

This initiative supports regional natural resource management groups; local, state and territory governments; Indigenous groups; industry bodies; land managers; farmers; Landcare groups; and communities.

Outlook

Climate Change

Australia's climate is changing as part of a global trend. Climate change is expected to continue and perhaps accelerate, presenting both challenges and opportunities for Australian agriculture. The effects of climate change will vary across sectors and regions, altering risk profiles both positively and negatively. Adapting to climate change will require information systems that identify the effects of climate change alongside other business management risks.

Agriculture is the dominant source of both methane and nitrous oxide emissions in Australia. Most agricultural emissions come from livestock (methane), burning of savannas and nitrous oxide emissions from soils.

Pests, Diseases and Weeds

More than 80 species of exotic vertebrate animals have established wild populations in Australia and more than 30 of these species have become agricultural or environmental pests. Major agricultural impacts of pest animals include:

- grazing and land degradation by rabbits and feral goats
- livestock predation by wild dogs, foxes and feral pigs
- damage to grain and fruit crops by mice and birds.

The direct costs to agriculture (including pest impacts and expenditure on management, administration and research) from wild dogs, rabbits, foxes, pigs, pest birds and mice was estimated to be approximately $745 million in 2009 (Department of Agriculture, Fisheries and Forestry [DAFF], 2012). Introduced insects, such as cattle ticks and aphids, cost agriculture more than $5 billion per year in production losses and $1 billion in control costs (DAFF, 2012).

Around 28,000 exotic plant species have been introduced into Australia since European settlement, and 3,480 of these have become weeds. Many of these weeds are escaped garden plants. Weeds contaminate crops, displace pasture plants and compete with crop and pasture plants for water and nutrients. Weeds also harbour diseases and insect pests, reduce livestock carrying capacity and condition and can be toxic to livestock.

The effect and control of weeds costs Australian agriculture more than $4 billion per year (DAFF, 2012). Farmers consider weed control one of their highest priorities in the prevention of long-term land degradation.

Biosecurity

Australia's biosecurity system minimises the risk of exotic pests and diseases entering the country and harming the natural environment. Australia's expanding economic ties with developing regions have led to inherent biosecurity risks. Warmer climates and faster transport systems can encourage pests and disease. Within Australia, disease and pest security is also increasing with climate change and improved logistics. The Australian Government is implementing reforms to Australia's biosecurity system to ensure it is responsive and targeted in a changing global trading environment.

Land Use

There is an overall trend of land use intensification in Australia, although this varies regionally and is set against a slow decline in the proportion of Australia's land area used for agriculture. According to ABS data, the area planted to crops (excluding pastures and grasses, and crops harvested for hay and seed) increased between 1992–93 and 2009–10, from almost 17.3 million hectares to almost 26 million hectares. Projected increases in population will create further pressure for land use intensification for residential, commercial and production purposes. The potential effect of climate change on agriculture and the possible contraction of food-producing areas, such as the Australian wheatbelt, will add to this pressure.

The causes and drivers of land use change are:

- pressures on resource availability, including land productivity, resource condition and population
- changing opportunities, including market development, production costs, new technologies, infrastructure and transport costs
- policy interventions, including subsidies, taxes, property rights, infrastructure and governance arrangements
- vulnerability and adaptive capacity, including exposure to natural hazards and the coping capacity of communities and individuals
- social changes, including changes in access to resources, income distribution and urban–rural interactions.

Land use change needs to be monitored to manage Australian landscapes and implement policy settings and program arrangements dealing with agricultural productivity, biosecurity, carbon, natural resources management, biodiversity and food security.

Australia is improving its capacity to track land use change, drawing on information such as satellite remote sensing and statistical collections. The Australian Bureau of Agricultural and Resource Economics and Sciences (ABARES) is working with the Australian Collaborative Land Use and Management Program's (ACLUMP) partners to promote collaboration among federal and state government agencies and others with interests in land use change analysis. The recent move by the Australian Government to establish a National Plan for Environmental Information, and by the ABS to introduce land and water accounting, will further promote tracking and reporting on land use change.

More recent statistical evidence from the ABS confirms this general pattern of change in agricultural land uses. Between 1992–93 and 2005–06, the area of agricultural holdings decreased by 5.5 per cent to 4,349,250 square kilometres. The most recent ABS information for 2009–10 shows that the area of land used for agriculture has continued to decline, to 3,985,800 square kilometres (a 13.4 per cent decrease from 1992–93). The area planted to crops (excluding pastures and grasses, and crops harvested for hay and seed) increased by 42 per cent between 1992–93 and 2005–06, and by 50 per cent between 1992–93 and 2009–10 (see Figure 3.7).

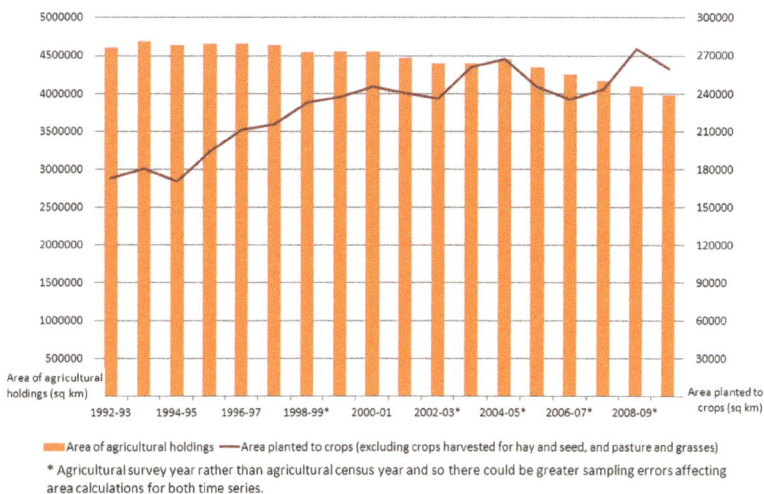

Area of agricultural holdings (sq km) — left axis; Area planted to crops (sq km) — right axis.

■ Area of agricultural holdings ── Area planted to crops (excluding crops harvested for hay and seed, and pasture and grasses)

* Agricultural survey year rather than agricultural census year and so there could be greater sampling errors affecting area calculations for both time series.

Figure 3.7: Agricultural land use change, 1992–93 to 2009–10.
Source: Lesslie, Mewett & Walcott (2011).

However, there is considerable variability in the spatial distribution of change across Australia over the period. For example, from 1993–94 to 2005–06, there was an increase in cropping area on the western slopes of New South Wales, western Victoria and generally across the grain-growing regions of South Australia and Western Australia (see Figure 3.8). There was a small decline in the area under cropping across most of northern New South Wales and southern Queensland.

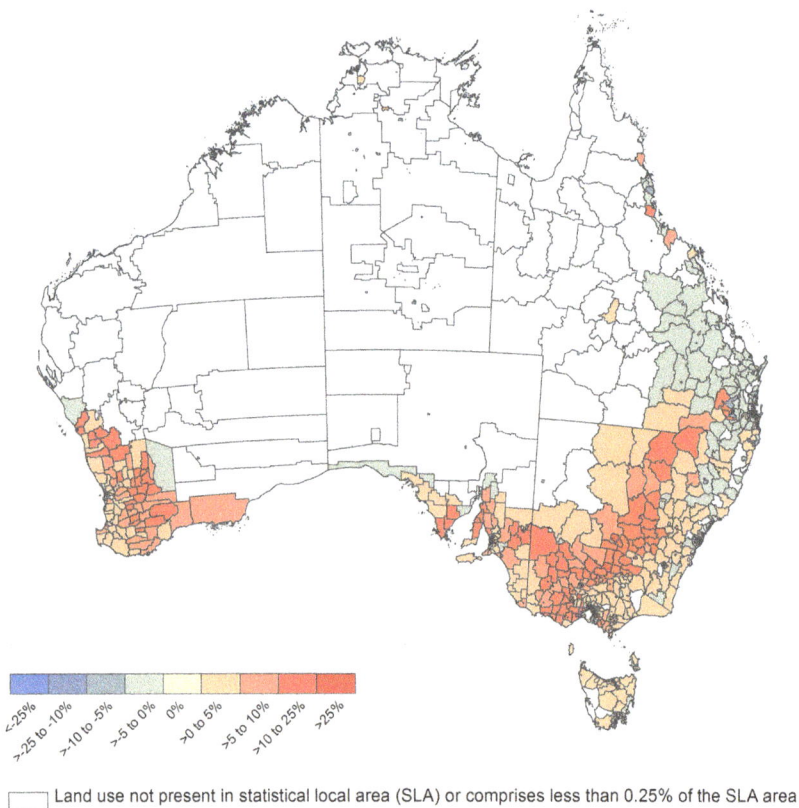

Land use not present in statistical local area (SLA) or comprises less than 0.25% of the SLA area

Figure 3.8: Change in area cropped, 1993–94 to 2005–06.
Source: Lesslie, Mewett & Walcott (2011).

In some cases, land use change can be represented as a process of land use intensification. Agricultural land use intensification is one response to the challenges of the cost-price squeeze faced by agricultural producers and increasing population. It reflects the attempt to secure more economic yield from each hectare through increasing concentrations of inputs, including nutrients, water, energy and management effort.

Agricultural land use intensification in Australia is illustrated in Figure 3.9, expressed as the cost of production per unit area. Intensification is generally concentrated in the more agriculturally productive regions that have a greater range of viable land use options, including opportunities for irrigation. Agricultural land use intensification is also concentrated in and around large population centres.

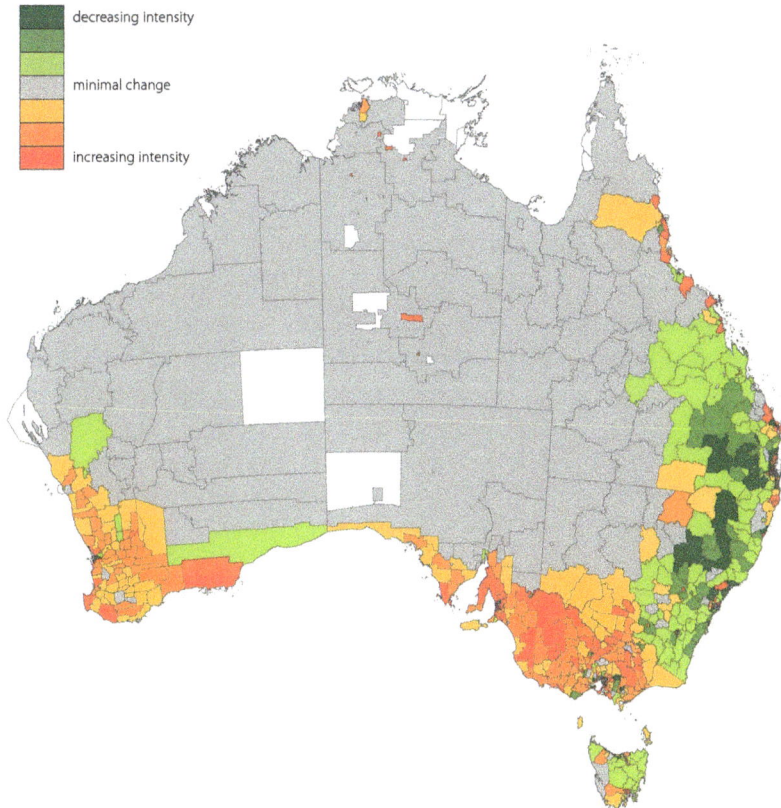

Figure 3.9: Change in agricultural land use intensification, 1985–86 to 2005–06.
Source: Lesslie, Mewett & Walcott (2011).

Table 3.3: Weighting factors

Land use category	Representative industry	Weighting factor
Residual or extensive grazing	Beef cattle	1
Sown pastures	Sheep/beef cattle	2
Grain crops	Grains	10
Irrigated pastures	Dairy cattle	40
Irrigated crops	Cotton, sugar cane, rice	50
Vegetables	Vegetables	80
Fruit	Fruit	80
Nurseries/turf	Horticulture	80

Note: Weighting factors for land use categories and representative industries used to calculate land use intensity index.

Source: Lesslie, Mewett & Walcott (2011).

Statistical collections from the ABS Agricultural Census and Agricultural Resource Management Surveys indicate that major changes in farm management practices are underway. For example, there has been a shift to conservation tillage over the past 15 years in broadacre cropping (see Figure 3.10). Conservation tillage helps promote improvements in soil carbon, reduced soil erosion and nutrient loss, cost savings and other production benefits. In 1996, conventional tillage (three or more cultivation passes) was the most common practice by area in all states except Western Australia and South Australia. By 2010, it was the least common practice in all states—'no cultivation' had replaced it as the most common cultivation practice.

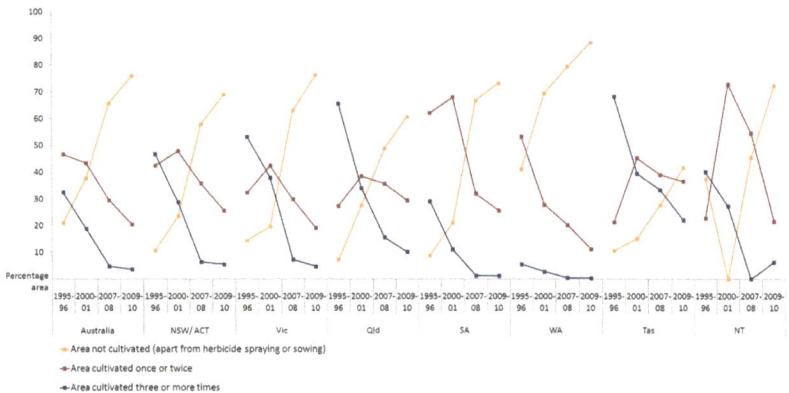

Figure 3.10: Cultivation practices for crops and pastures, 1995–96 to 2009–10.

Source: ABS (2010a, 2010b).

Ecosystem Services

An ecosystem services approach integrates the ecological, social and economic dimensions of natural resources management, including conservation and production objectives, by explicitly identifying and classifying the benefits from ecosystems. These include market and non-market, use and non-use, and tangible and intangible benefits.

The ecosystem services concept has been used successfully in Australia and internationally to identify natural resource management priorities at the catchment, regional, national and global scales, and to report on the relationship between the environment and human wellbeing. For nearly two decades, it has been a component of Australian Government policies and programs focused on sustainable management and development. For example, ecosystem services were a feature of the policies and programs that flowed from the National Strategy for Ecologically Sustainable Development, especially the Natural Heritage Trust and the National Action Plan for Salinity and Water Quality. The ecosystem services concept is also identified in the overarching goal of the Caring for our Country initiative.

The ecosystem services approach is particularly relevant to using and managing land cover. Land cover is related to a range of ecosystem services, such as helping to regulate water flow and maintain water quality. Such services have traditionally been treated as public goods with little or no explicit financial value, but this is changing. Potential payments for vegetation-based services, such as biodiversity conservation, carbon sequestration, salinity abatement and opportunities for ecotourism, wildlife photography and environmental education, can provide significant environmental, economic and social benefits and contribute to reducing the cost of management. While markets for such services remain a minor component of the national economy, they are expected to grow.

Research Capacity — Land Use and Management

ABARES provides professionally independent economic and scientific analysis, including integrated socio-economic and biophysical analysis that informs the difficult policy issues facing Australia's primary industries.

It has staff with skills across a range of economic, science and social science disciplines. These include scientific and economic analysis and modelling; data (including survey) collection and statistical analysis; risk assessment and management; geographical mapping, particularly in areas related to natural resource management; commodity and market analysis; and integrated analysis.

ABARES has an established capacity to compile national land use, land management and land cover information, and track change using regular statistical collections by government and industry and remotely sensed imagery. These types of information are combined to take advantage of their complementary spatial and temporal characteristics—for example, in the national-scale land use mapping produced by ABARES.

Tracking and Reporting Change

ABARES is working with ACLUMP partners to promote collaboration among state and federal government agencies and others with interests in land use change analysis. Australia's large land area means that remote sensing is an attractive option for cost-effective mapping of aspects of land use change. Free access to imagery archives, such as Landsat, MODIS (Moderate Resolution Imaging Spectroradiometer) and AVHRR (Advanced Very High Resolution Radiometer), has resulted in ready uptake in mapping programs. Agricultural statistics compiled by the ABS and ABARES are also important sources of information for analysing and reporting change.

National Land Use Mapping

National-scale land use mapping is modelled using coarse-scale satellite data (pixel size of 1.1 square kilometres): ABS Agricultural Commodity Census statistics for agricultural land uses; pre-existing finer resolution data (principally at the 1:250,000 scale) for other uses. The relatively low cost of national mapping provides an opportunity for time-series mapping. National-scale (1:2,500,000) datasets have been completed for 1992–93, 1993–94, 1996–97, 1998–99, 2000–01, 2001–02 and 2005–06 (see Figure 3.1 for the most recent map). The next national-scale dataset will be based on the 2010–11 Agricultural Commodity

Census. National-scale mapping produced by ABARES is in strong demand for synoptic-level land use assessments, and for strategic planning and evaluation (such as developing programs for natural resource management). It is also used in modelling applications, such as national carbon accounting and salinity assessments, at the river-basin level.

Dynamic Land Cover Mapping

A dynamic land cover map and databases for Australia produced by Geoscience Australia in partnership with ABARES provide new insights into aspects of land use and land management change for Australia. The map and time-series data, produced for 2000–08 by ABARES using MODIS EVI (Enhanced Vegetation Index) data, allow trends and changes in land cover over time to be investigated. This database can provide insight into the response of land cover to a wide variety of drivers—both natural and anthropogenic. This allows natural resource managers to identify emerging patterns of land cover change, and provides a broad spatial and historical context within which to interpret that land cover change.

Ground Cover Monitoring for Australia

Ground Cover Monitoring for Australia is a national program, coordinated by ABARES, that involves the remote sensing of fractional cover (i.e. green cover, dry cover and bare ground) across Australia using MODIS NBAR (Nadir Bi-directional Reflectance Distribution Function Adjusted Reflectance) data. The program includes the establishment of a national system of ground validation sites using nationally agreed methods. The data are being used initially to support soil erosion modelling, and it is intended that this will be extended to mapping management practices (such as tillage and stubble management) in the cropping zone. Partners include key national research organisations (i.e. Commonwealth Scientific and Industrial Research Organisation [CSIRO], Geoscience Australia and Terrestrial Ecosystems Research Network) and natural resources management or agriculture departments in each state.

Land Management Practices

Land management practices are analysed primarily using agricultural statistics collections, such as those of the ABS and ABARES. In 2007–08, the Agricultural Resource Management Survey was conducted by the ABS to provide a baseline of key practices to help guide national investment programs to improve natural resources management. Practices include those relating to tillage, stubble management, ground cover management, fertiliser use, soil testing and liming. This survey was also run for 2009–10, and will be run again for 2011–12. These results will be used to measure change in land management practices over time.

ABARES carries out a smaller annual survey—the Australian Agricultural and Grazing Industry Survey—of the broadacre cropping, grazing and dairy industries. This survey produces estimates of land area and tenure, labour, farm capital, crop type and production, fertiliser use, irrigation, farm receipts, farm costs, farm performance measures, farm debt and farm equity.

Spatial Decision Support

The Multi-Criteria Analysis Shell for Spatial Decision Support (MCAS-S), developed by ABARES, is an easy-to-use spatial decision support tool designed to help visualise and combine mapped information in a flexible, interactive way. MCAS-S is the latest of several multi-criteria decision support aids used by DAFF since the early 1990s to support policy decision-making.

MCAS-S allows the user to import, select and display spatial data in a dynamic workspace window, see multiple datasets simultaneously, group datasets under themes, interactively modify and combine datasets, and carry out two-way and multi-way comparisons to form meaningful map-based flow diagrams. Layers can be combined using simple weights, complex functions or through pair-wise comparison. It also allows users to document their results and the decision-making process, including assumptions.

A project can be constructed at any scale and resolution and 'live updates' are available, which are particularly helpful at workshops. MCAS-S assists in decision-making in situations in which transparency between different approaches to map combinations is needed. Stakeholders can see the effects their decisions may make. Successful use of the software does not require geographic information system (GIS) programming, which removes the usual technical obstacles non-GIS users have in accessing and analysing spatial information.

MCAS-S is being used at the national, regional and catchment scale for:

- wind erosion extent and risk assessment, 2009
- soil acidification risk assessment, 2010
- soil carbon potential evaluation, 2009
- targeting investment—Great Barrier Reef water quality, 2009–10
- indicators of community vulnerability and adaptive capacity, 2010
- rabies risk mapping, 2011
- weeds risk assessment, 2011
- Asian honey bee risk assessment, 2009
- animal disease risk mapping, 2011
- revegetation planning, 2006
- land acquisition priorities for conservation, 2010
- catchment revegetation planning, 2011
- priorities for regional natural resources investment, 2011
- wildfire assessment (soil erosion), 2011
- agricultural land quality evaluation, 2011.

Figure 3.11 shows data layers that have been combined using MCAS-S to produce a national map of the extent and severity of wind erosion. These data included modelled wind erosion data (created using a model), an index of dust storm activity based on observed data, and expert opinion (rankings by region) of the extent and severity of wind erosion. The input data layers were weighted for each state according to expert opinion on their confidence in each layer, and were combined spatially using the MCAS-S tool. The darker areas are those with the highest wind erosion extent and severity in Australia, as determined by an expert panel.

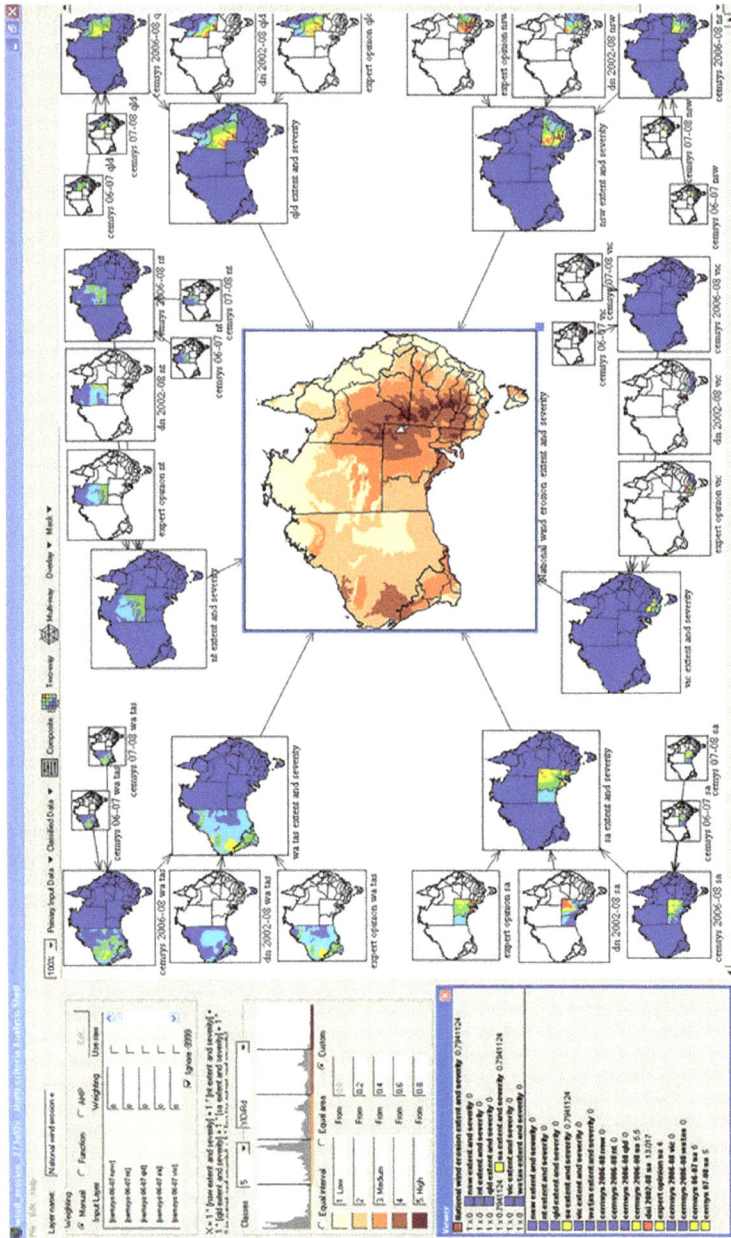

Figure 3.11: Multi-Criteria Analysis Shell for Spatial Decision Support project screenshot to determine the extent and severity of wind erosion in Australia.

Note: The states were analysed individually and then combined to provide a national picture.

Source: Smith & Leys (2009).

Key Issues

This chapter was originally prepared as a background paper as part of a collaborative project between ABARES and the Forestry Economics and Development Research Center of China's State Forestry Administration to develop a sustainable land and forest management research agenda.

Several potential issues in the land use and resources area were identified as context for discussions between ABARES and the China National Forestry Economics and Development Research Center on common technical interests. These include:

- Australia's food supply system and its potential to contribute to food security domestically and globally
- where and how sustainable agriculture can be maintained and developed in response to climate change and loss of productive agricultural land
- options for managing carbon in Australian agricultural production systems (including farm forestry)
- tracking 'hotspots' of land use change, including land use intensification on the fringes of cities and urban areas, and loss of productive agricultural land, productive land resources and biodiversity
- transitions in irrigated agriculture, forestry and carbon farming, and the trade-offs between biofuels and other forms of agriculture.

Addressing these issues will require:

- more effective use of newly available remote sensing and statistical information
- new methods for change analysis (including automated change detection), and better measurement of error and uncertainty
- better links between existing data collection and analysis activities, including international engagement
- better methods for identifying and classifying thematic transitions, cyclic variability and trends
- better methods for forecasting land use change, including change in response to increased climate variability and disaster risk (e.g. biosecurity, fire, flood and cyclone)

- land use and land management incentives for non-market ecosystem goods and services
- decision support capability to enable informed, transparent consideration of options and trade-offs.

Potential Areas for Collaboration

The Sustainable Land and Forest Management Research Agenda project has strengthened technical cooperation in areas of common interest. To further improve the sustainable management of land and forests, potential areas for collaboration have been identified. These include:

- spatial decision support systems
- land resources assessment
- landscape classification
- ecosystem services
- remote sensing of land use and land cover change
- climate change adaptation
- carbon accounting relating to land use and land management change.

References

Australian Bureau of Agricultural and Resource Economics and Sciences. (2011). *Guidelines for land use mapping in Australia: Principles, procedures and definitions.* A technical handbook supporting the Australian Collaborative Land Use and Management Program (4th ed.). Canberra, ACT: Australian Bureau of Agricultural and Resource Economics and Sciences.

Australian Bureau of Agricultural and Resource Economics and Sciences – Bureau of Rural Sciences. (2010). *Land use of Australia, 2005–06* (version 4). Canberra, ACT: Australian Bureau of Agricultural and Resource Economics – Bureau of Rural Sciences.

ABS (Australian Bureau of Statistics). (2008). *Agricultural resource management survey 2007–08.* Retrieved from www.abs.gov.au/ AUSSTATS/abs@.nsf/DetailsPage/7101.0Mar%202008

ABS. (2009). *Land management and farming in Australia 2007–08* (Report no. 4627.0). Canberra, ACT: Australian Bureau of Statistics.

ABS. (2010a). *Agricultural census, 1995–96 and 2000–10.* Retrieved from www.abs.gov.au/AUSSTATS/abs@.nsf/Lookup/4627.0Main+ Features12011-12

ABS. (2010b). *Agricultural resource management surveys, 2007–08 and 2009–10.* Retrieved from www.abs.gov.au/AUSSTATS/abs@.nsf/ Lookup/4653.0Main%20Features12Dec+2008

Bureau of Rural Sciences. (2006). *Guidelines for land use mapping in Australia: Principles, procedures and definitions. A technical handbook supporting the Australian Collaborative Land Use Mapping Programme* (3rd ed.). Canberra, ACT: Bureau of Rural Sciences.

Clancy, T. & Howell, C. (2013). *Sustainable forest management: The Australian context.* Australian Bureau of Agricultural and Resource Economics and Sciences (Research Report 13.2). Canberra, ACT: Australian Bureau of Agricultural and Resource Economics and Sciences.

DAFF (Department of Agriculture, Fisheries and Forestry). (2008). *Australia's agricultural industries at a glance.* Canberra, ACT: Department of Agriculture, Fisheries and Forestry.

DAFF. (2012). *Australia's agriculture, fisheries and forestry at a glance.* Canberra, ACT: Department of Agriculture, Fisheries and Forestry.

Gammage, B. (2011). *The biggest estate on earth.* Sydney, NSW: Allen & Unwin.

Lesslie, R., Mewett, J. & Walcott, J. (2011). *Landscapes in transition: Tracking land use change in Australia. Science and Economic Insights, 2.2.* Canberra, ACT: Australian Bureau of Agricultural and Resource Economics and Sciences.

McKenzie, N., Jacquier, D., Isbell, R. & Brown, K. (2004). *Australian soils and landscapes: An illustrated compendium.* Melbourne, VIC: CSIRO Publishing.

Smith, J. & Leys, J. (2009). *Identification of areas within Australia for reducing soil loss by wind erosion.* Canberra, ACT: Bureau of Rural Sciences.

4

Does Public Policy Obey Data, Information and Maps?

Stephen Dovers

Key Points

- The link between science (maps and data) and policy is complex and variable: rarely strong and direct; sometimes discernible and positive; often hard to discern or sadly missing. Other things matter as well in political decisions.
- Australia has enviable capacities in land use science, as this volume shows, and although the application of this science may disappoint, our situation would be far worse without this excellent knowledge base.
- Lack of continuity of data and initiatives is a major issue for rational and sustainable land use decision making.
- A better understanding of science by policy makers, and of the realities of policy and politics by scientists, could improve the situation.

Endless data, information and maps instruct us that Australia—and indeed the world—should be doing considerably more than is being done to ensure the sustainable use of natural resources, both in the interests of the inherent value of natural ecosystems and the future wellbeing of human populations (e.g. Lindenmayer et al., 2014; United Nations, 2012; United Nations Environment Programs, 2012). The information base is not equivocal—species and ecosystems are being lost at an alarming rate, the natural resource base of too many societies is being degraded at an unsustainable rate, the stability of climate is being disturbed, and the capacity of natural systems to absorb our waste is regularly exceeded.

As the chapters in this volume show, Australia is a leader in environmental data gathering. It is also a leader in the transformation of such data into models, maps and indicators that are suited to use by land managers and policy agencies. Yet, the scientists involved generally regard the use and effect of this knowledge base as insufficient.

Why don't local, national and international policymakers obey the messages from 'science',[1] implement appropriate policies in a rational manner and address the problems more purposefully? Some people, including many scientists and economists, believe that politics and policy should be entirely rational, taking on board expert evidence and enacting policies that address manifest problems. In the discipline and practice of public policy, this fantasy was abandoned long ago (see Dovers & Hussey, 2013). The processes that determine public policy are not, and can never be, linear and rational. There is too much at stake and too many competing interests to be considered: ecosystem integrity, poverty alleviation, maintenance of employment, social values and public support, international policy commitments, competition for available funding, public policy effort and more. Of course, politics can be messy, bloody, divisive and myopic. For the moment, we can take the scientific and official international policy consensus that the current answer is not nearly enough.

There are counter cases, as evidenced by the many decisions made to protect the environment and to use natural resources more judiciously— policy decisions about pollution, endangered species, water abstraction, ozone-depleting substances and so on. Australia's protected areas network has increased by an order of magnitude over the course of the last century; this has been informed by ecological science findings describing the decline of native species and loss of ecosystems, among other influences. Similarly, Australia's world-leading experiments in connectivity conservation owe much to Australia's world-leading landscape ecology. The landmark 1975–77 collaborative federal–state soil conservation

---

1   I use the term 'science' broadly to mean the information emerging from the natural sciences (ecology, physics, hydrology, etc.), but also social science disciplines (demography, economics, sociology, etc.). When differentiation is needed, it will be apparent.

study, and its later transformation into maps (see Woods, 1983), was fundamental to the Decade of Landcare initiative and what followed in national policy. However, science never operates alone. That land degradation data and mapping exercise was important, but so too was the genius of an alliance between the National Farmers Federation and the Australian Conservation Foundation. Led by Rick Farley and Phillip Toyne, it had the support of prime minister Robert Hawke, an advocate of consensus who was ready to seize the moment. The dramatic erosion-fuelled dust storm over Melbourne that filled the television screens of urban Australians (who were usually unaware of non-metropolitan issues) was also important, as was the experience of programs and practices built up over decades by the various state soil conservation agencies. Science, politics, practical agency knowledge and opportune media moments combined to create interest in the environment.

Why has environmental action not followed (or been perceived to follow) environmental science more closely? The answer is at once simple and highly complex. The simple answer is that 'science'—environmental or otherwise—is only one of several information inputs to policy: other messages matter too, whether one agrees that they should or not. The complex answer is that the story of how policy is made and implemented, and what information and imperatives influence it, varies enormously from issue to issue, jurisdiction to jurisdiction and over time. There are no strict rules or predictable patterns. To make some sense of this complexity within the context of this volume, this chapter briefly presents some key concepts and frameworks from the discipline of public policy, illustrated with examples from the natural resources and environment domains. This will be done in two parts: evidence-based policy and information in policy.

What is the 'Evidence' in Evidence-Based Policy?

There is an expectation that the policy decisions of governments, made after consultation and interaction with partners from the private and community sectors, will be based on solid, rigorous information and analysis. In recent years, the term (and aspiration of) 'evidence-based' policy has become commonplace. This raises an important question:

what 'evidence' is required to fulfil the modern promise of evidence-based policy? Head (2008) defined three forms of evidence that are called upon when policy is made:

1. systematic ('scientific') research
2. program management experience ('practice')
3. political judgement.

This simple but defensible schema places environmental science in context; and it certainly played out in the soil and land degradation case described above. In the domain of land use and environmental management, 'scientific' research and information is not only one of three forms of evidence, but also includes more than ecology and landscape science and the like—it also incorporates economics, sociology, demography and more. The 'scientific' consensus of what should be done—assuming there is one (often there is not)—may be rather divided depending on what value or asset, and thus what discipline, is the priority. Can we quantitatively and scientifically trade-off threatened species, groundwater aquifers, rural employment and legal obligations under international trade and other agreements? This is when the other two categories of evidence come into play: policymakers and advisors must determine whether a path of action is practically achievable or advisable in a public policy sense, and whether governments and the public are likely to accept it. Davis, Wanna, Warhurst and Weller (1993) summed up the political reality:

> Politics is the essential ingredient for producing workable policies, which are more publicly accountable and politically justifiable ... While some are uncomfortable with the notion that politics can enhance rational decision-making, preferring to see politics as expediency, it is integral to the process of securing defensible outcomes. We are unable to combine values, interests and resources in ways which are not political. (p. 257)

Policy directions are set by governments; they reflect an interpretation of social goals and always involve 'value, interests and resources'—matters that are beyond linear, rational scientific answers. Policymaking is a social and political process; like politics itself, it is sometimes short-sighted, combative and disorderly. For example, after an unprecedentedly rigorous investigation, the Resource Assessment Commission (RAC) reported that a proposed uranium mine at Coronation Hill was economically valuable, environmentally problematic (but manageable) and damaging to Indigenous culture (Stewart & McColl, 1994). Whether the mine should go ahead could not be quantified or rendered to a composite metric that provided the yes-or-no answer wanted by some, for this required a value judgement—a *political* judgement. While research—scientific inquiry, data gathering and analysis—could enable and inform that judgement, it was up to politicians to make it—specifically, federal cabinet. Internationally remarkable, the RAC linked expertise and evidence with public policymaking and broke new ground in methods and processes; however, it lasted only four years. Embroiled in near-term, politically sensitive issues, its rigorous 'scientific' interpretations probably contributed to its demise. Science (and economics and social science, in the Coronation Hill case) can get too close to policy and politics.

That leads to the first, crucial point of this chapter. In a liberal democracy, policy is not (and never will or should be) made by 'experts'; rather, it is made by politicians and governments, reflecting social values and aspirations. Rob Lesslie's foundational work on wilderness (see Chapter 1) did not *make* policy: societal values had shifted towards valuing wilderness, and government wished to follow this shift. Rob's work provided a robust evidence base from which sound decisions could be made.

How Does Policy Use Information?

Confining ourselves to Head's 'systematic research' as an input to policy, we can consider different ways in which information or evidence might be used in policy. Sometimes direct utilisation and impact occurs, leading to a discernible policy change. However, this does not occur very often, for not only do other forms of 'evidence' matter, but information can also be put to positive use, used negatively or simply ignored. Applying long-discussed (and contested) concepts from theoretical and practical public policy literature to composite environment and sustainability indicators, Hezri (2004) identified five forms of information utilisation in policy: instrumental, conceptual, symbolic, tactical and political. He mapped these against the nature of the response to information and the degree of rationality of use, as shown in Table 4.1.

Table 4.1: Taxonomy of indicator (information) use.

Nature of response	High degree of rationality	Low degree of rationality
Positive	Instrumental use	Political use
	Accepted and used to direct or inform policy choice or design.	Used to support a predetermined policy position, not necessarily with any concern over quality.
Ordinary	Conceptual use	Symbolic use
	Information 'seeps into' policy discourse, reframing, with a subtle impact over time. Can define the policy agenda.	Used to assure other parties that something is being done; the situation is being considered.
Negative	Not used	Tactical use
	Information discarded or disregarded.	Incomplete information, or further information gathering, used to delay decisions, as a substitute for action, or to deflect criticism.

Source: Adapted from Hezri (2004).

Consider any major contested issue, past or present, and information will have been 'used' in more than one of the ways defined in Table 4.1: to support or oppose some policy action, or to deflect criticism or defer a decision. It may have been contested; it may have been ignored. While evidence of clear instrumental use is unusual, the argument for conceptual use is much stronger; however, it is a diffuse phenomenon to trace and attribute.

Instrumental or direct use should not be overinterpreted. It is very rare that science directly drives or determines research, but it may well be an important reference point or input (e.g. Rob Lesslie's wilderness work). Major scientific efforts have been behind the setting of sustainable diversion limits and environmental water allocations in the Murray–Darling Basin; however, expert consensus did not determine the final number: the science informed, but did not determine the policy outcome. Is 2,750 gigalitres too little or too much? The answer is a matter of science; it is also a matter of normative values and competing imperatives.

Science can influence policy agendas in different ways and sometimes gains purchase by narrowing the focus of policy attention. The integration of an array of issues—including soil erosion, rangeland vegetation decline, acidification, soil structural decline and salinity—into the one agenda of land degradation by Woods (1983) and others was a major advance in understanding and conceptualisation. However, driven by environmental scientists, this agenda was quickly dominated by dryland salinity. Acidification, erosion and structural decline were still there, but we stopped talking about them, and funding and policy faltered. Perhaps we can only twiddle one knob at a time? In the 1980s, Australia began to embrace the United States' lead and agenda regarding 'instream flows', a package of non-extractive values of water in situ in rivers, including ecological, recreational, aesthetic and cultural values, and geomorphic integrity. The Australian discourse and policy direction restricted rapidly towards ecological values and adopted the term 'environmental flows'; this move reduced the evidentiary power of non-extractive water values and alienated those for whom an 'environmental' agenda was not attractive.

When 'science' sets the policy agenda, or has a distinct impact, it may be that only one set of opinions, disciplines or findings is in play. Scientists from the same discipline or even subdiscipline are often called upon as expert witnesses by opposing sides of legal battles, and disagree superbly. Which 'scientist' should a policymaker or a judge believe, and why? What standard of proof must they display to earn such belief?

Uncertainty and Standards of Proof

Arguably, the natural sciences are more cautious in the face of uncertainty than others engaged in policy debates: 95 per cent confidence limits are a high standard, and anonymous peer review is, for all its occasional faults, a rigorous process. In some science-to-policy instances, there is an agreed nomenclature describing uncertainty and confidence (the nomenclature used by the Intergovernmental Panel on Climate Change being the most famous), but often there is not. However, different confidence limits used in science are only part of the story. Does 95 per cent equate to the criminal law's 'beyond reasonable doubt'? Is it certainly higher as a standard of evidence than the civil law's 'on the balance of probabilities'? Does the civil law standard equate to the public policy theory and practice of 'satisficing'? How does a seasoned policy official assess the probability of gaining approval for a proposal from the relevant minister and cabinet? While such an assessment may draw on 'systematic (scientific) research', being the first of Head's three lenses, it would rely more heavily on 'program management experience' and 'political judgement'. What standard of proof is used, either implicitly or explicitly, by other actors in the environment domain: the media, consultants doing an environmental impact statement, industry and union officials defending jobs, professional environmental lobbyists, or local residents opposed to coal seam gas developments? Two real but anonymised exchanges illustrate the difference:

1. In a discussion around identifying wildlife corridor (connectivity conservation) projects that are of significance:

Senior scientist: 'I'm wary of lines on the maps ... the uncertainties around definite boundaries and around which projects would be the most effective for conservation are too great'.

Senior government official: 'Well if you won't I'll get a bloody big thick texta and do it myself'.

2. Presenting a predictive urban land surface mapping method drawn from remote-sensed data:

Spatial scientist: 'This hasn't really worked ... ground truthing is only getting us to about 80 per cent accuracy'.

Urban planner: '80 per cent: that's luxury! Anything near 60 and we take it on board in future planning'.

A lack of understanding of different standards of proof and ways of presenting evidence is rife. Many people outside research do not understand the scientific method, confidence limits or peer review. Equally, many scientists (and social scientists) do not understand how public administration, public policy, the courts and politics function. The situation is made worse by tactical, symbolic and political uses of information in public debates, and by the inability (or refusal) of the media to convey complexity, uncertainty and the nature of 'systematic (scientific) research'. There is much discussion of how mutual understanding could be improved: science communication, cross-institutional secondments, citizen science, new forms of media (e.g. *The Conversation*), knowledge brokers and boundary organisations, and whole journals devoted to the interface (e.g. *Environmental Science and Policy*). However, for those scientists close to the game and concerned with engaging with policy, I have a simpler and quicker suggestion: read a policy textbook and at least know some of the landscape and terminology—I recommend Dovers and Hussey (2013).

A final observation on uncertainty (or an indulgent diversion to make a point). There is, of course, no standard of proof or evidence rule that can be applied to policy proposals; they vary too much in their substance and implications. To give a crude illustration: a land use or production change driven by strong evidence of environmental damage that would lead to 50 job losses will always be judged differently to one that would cause 500 job losses. The stakes matter and the trade-offs will always be evaluated through the policy and political process. The precautionary principle instructs care in the face of 'serious or irreversible' environmental harm; this is something we can generally comprehend—loss of topsoil, extinction of a species or poisoning a slow-moving aquifer. Can we turn the 'irreversibility' test against policy as part of the conversation around standards of proof and trade-offs (Dovers, 2006)? Just as some environmental damage cannot be reversed, the same can also be said for policy decisions: consider privatising a natural resource management function, changing property rights or discontinuing, for even a few years, an environmental data time series. Given that it is extremely rare that a policy can be guaranteed to have a specific impact, thus 'satisficing' a reasonable standard of proof or evidence, should a policy or institutional change that cannot be (or cannot easily be) reversed be subject to a higher standard—perhaps even beyond reasonable doubt?

Conclusion

Does policy obey data, information and maps? This brief chapter repeats the standard public policy answer: sometimes a bit, sometimes a lot and sometimes not at all. This may be an unsatisfying answer for many people who wish for rational, science-based decisions to be standard; however, hopefully, the answer is at least more comprehensible when the nature of evidence and information use in policy is explained.

To end on an untestable but mildly optimistic point: Australia is home to world-class environmental and landscape science; scientists are often engaged closely with policy and land management agencies and there is some evidence of their impact. They may be frustrated that not all decisions and practices are consistent with their evidence, but they might want to consider how Australia's natural resources and ecosystems would be faring without them.

References

Davis, G., Wanna, J., Warhurst, J. & Weller, P. (1993). *Public policy in Australia*. Sydney, NSW: Allen & Unwin.

Dovers, S. (2006). Precautionary policy assessment for sustainability. In E. Fisher, J. Jones & R. von Schomberg (Eds.), *Implementing the precautionary principle: Perspectives and prospects* (pp. 88–109). Cheltenham, UK: Edward Elgar. doi.org/10.4337/9781847201676. 00014

Dovers, S. & Hussey, K. (2013). *Environment and sustainability: A policy handbook* (2nd ed.). Sydney, NSW: Federation Press.

Head, B. (2008). Three lenses of evidence-based policy. *Australian Journal of Public Administration 67,* 1–11. doi.org/10.1111/j.1467-8500.2007.00564.x

Hezri, A. A. (2004). Sustainability indicator system and policy processes in Malaysia: A framework for utilisation and learning. *Journal of Environmental Management 73*, 357–71. doi.org/10.1016/j. jenvman.2004.07.010

Lindenmayer, D. B., Dovers, S. & Morton, S. (Eds.). (2014). *Ten commitments revisited* (2nd ed.). Melbourne, VIC: CSIRO Publishing.

Stewart, D. & McColl, G. (1994). The Resource Assessment Commission: An inside assessment. *Australian Journal of Environmental Management 1,* 12–23.

United Nations. (2012). *The future we want: Rio +20 United Nations conference on sustainable development, outcomes of the conference.* New York, NY: United Nations. Retrieved from www.un.org/en/ sustainablefuture/

United Nations Environment Programs. (2012). *Global Environment Outlook GEO5: Environment for the future we want.* Nairobi, Kenya: United Nations Environment Program.

Woods, L. E. (1983). *Land degradation in Australia.* Canberra, ACT: Australian Government Publishing Service.

Part 2 – Ad Hoc or Strategic Responses

5

Addressing a Lost Opportunity: Towards Science-Informed Land Use Planning

Darryl Low Choy

Key Points

- Post–World War II (WWII), Australia was a leader in the development of scientific information systems to support natural resource management including the assessment of a landscape's non-urban development potential.

- Surprisingly, there was little uptake of these environmental resource mapping initiatives into the land use planning sector, which deals with significant landscape modifications.

- Overlooking of this noteworthy development represented a major lost opportunity to better inform the decision-making process through a land use planning process that was based on reliable biophysical science.

- Alternative planning paradigms have subsequently emerged that address this lost opportunity and demonstrate the use of science to better inform the land use planning process.

- Holistic and environmentally based planning, with a strong nexus between land use planning and environmental sciences, will require a 'marriage' between planners and scientists.

Introduction

Australia has led the way for many decades in the design and development of scientific information systems to support natural resource management. It was an early innovator in the use and application of aerial photography (principally vertical, but also oblique) for the purposes of landscape interpretation. However, there has been surprisingly little uptake of these systems and their content into the land use planning sector, in which most landscape-scale decisions that affect the environment and can lead to significant landscape modifications are made. Land use planning refers to various forms of statutory, but also non-statutory, planning. This is referred to in the literature under a variety of terms, such as town, city, spatial, physical or urban and regional planning. It is the planning that is undertaken by local government and certain state government agencies over freehold and crown lands for the purposes of defining the future location and scale of urban and rural settlements and land use activities.

The early environmental (landscape) resource mapping initiatives were never intended to be used to support land use planning endeavours; however, in situations in which they coincided geographically, they could have provided an excellent foundation upon which plans for the future development of our cities, especially the regions, might have been based. In other circumstances, their methodology could have been employed to better inform the planning process. The land use planning sector has an extremely poor track record of utilising these scientifically based information resources; disconnections between scientific information systems and regional-scale planning initiatives have dominated national and state government policy at significant times in Australia's development. This chapter provides an overview of selected major regional-scale Australian land use planning initiatives to identify their degree of nexus with parallel scientific information systems of similar scale and focus. Some consideration of the circumstances of these relationships and lessons for the future are offered by way of conclusion.

Post-WWII Resource Mapping (1946–70s)

WWII highlighted the paucity of adequate mapping across the continent, as well as the sparsely populated and unproductive nature of its northern regions. Rapid topographic mapping was commenced in 1942 utilising trimetrogon reconnaissance photography. By the end of the war, vertical aerial photography had become the major tool of cartographers. Number 87 Squadron of the Royal Australian Air Force was the principal vertical aerial photography agent for the national mapping effort until it was disbanded in 1953. This extensive and systematic vertical aerial photography effort led to a complete coverage of the Australian continent by the mid-1960s (with photos at nominal scales of 1:50,000 to 1:84,000). Overseen by the National Mapping Council, these strategic base mapping initiatives were coordinated by the National Mapping Office (from 1951), then the National Mapping Section, Property and Survey Branch, Department of Interior (later, the Division of National Mapping). These early national mapping initiatives left a rich legacy of one-inch to the mile and 1:250,000 topographic maps supported by continental coverage of black-and-white vertical aerial photographs (largely at approximately 40,000-feet flying height).

In the post-WWII era, the Australian, Queensland and Western Australian governments prioritised the need to undertake systematic surveys of the land resources and development potential of northern Australia. The Commonwealth Scientific and Industrial Research Organisation (CSIRO) was tasked to undertake a reconnaissance survey of the Katherine–Darwin area in 1946 that culminated in the publication of the first report in CSIRO's Land Research Series (Christian & Stewart, 1952). By 1977, when the series ended, it had produced 39 reports across 36 regions, including 17 reports dealing with regions in Papua New Guinea. CSIRO's initiative required a new approach to integrating biophysical data, including topography, vegetation and soils, and underlying causal factors such as climate, geology and geomorphology. It was argued that:

> This information had to be collated and mapped at a scale that was both appropriate for spatial description of agricultural, pastoral and settlement potential and relevant to the needs of policy-makers. The report also needed to identify priority areas and possible approaches to development of the area's land resources. (CSIRO Publishing, n.d.)

The resultant composite mapping unit was named the 'land system'; this was defined as 'an area or group of areas throughout which there is a recurring pattern of topography, soils, and vegetation' (Christian & Stewart, 1952). This approach relied on the mapping of landscape patterns that could be identified from aerial photography or photo-mosaics supported by topographic and thematic maps, reports, fieldwork and expert knowledge. Aided by WWII-perfected stereoscopic aerial interpretation techniques, the continental coverage of vertical aerial photographs greatly facilitated this land research mapping initiative.

The late 1960s saw another CSIRO division (soil mechanics) develop a further specific application of terrain evaluation. This initiative also utilised stereoscopic aerial photograph interpretation techniques. This system of terrain evaluation was known as the 'PUCE' (Pattern-Unit-Component-Evaluation) program. It was based on the acknowledgement and recognition that four parameters defined an area: the underlying geology, slope, soil and vegetation characteristics. The PUCE program provided a reliable and rapid method of evaluating the landscape in terms of its recurring composite patterns, units and components. As in the case of the early Land Research Series, the new initiative, known as Terrain Evaluation for Engineering Purposes, sought to 'provide rational formalized systems for … planning, engineering works on a regional basis' (Grant, 1971, p. 81). By the late 1970s, large regional areas of northern and Central Australia had been mapped, as well as the principal metropolitan regions of Brisbane, Sydney, Melbourne and Perth and the nominated growth area of Albury–Wodonga.

Post-WWII Planning (1946–50s)

The post-WWII era witnessed the delineation of reconstruction regions that were to involve large national-scale initiatives to 'secure a peace time economy of full employment … [in which] housing and town planning were seen as crucial to raising living standards' (Freestone, 2012, p. 1). This initiative was overseen by the Australian Government's Department of Post-War Reconstruction (1942–50). It was noted that some relevant recurrent themes included the importance of a multidisciplinary approach across physical, economic and social ends; coordination across all three levels of government; treating the urban environment holistically; and research-based evidence and scientific values. Citing H. C. Coombs, an influential bureaucrat of this era, Freestone observed that the Department of Post-War Reconstruction was 'concerned with the physical aspects of planning and development strategy as a countervailing force to dependence on purely financial instruments of resource allocation' (p. 4). However, these national-scale planning proposals took a back seat with the realisation that it was the states that had constitutional responsibility for physical land use planning, and that many of these proposals could be viewed as federal intrusion into state matters.

In parallel with this (largely non-spatial) post-WWII reconstruction planning, a number of metropolitan- and regional-scale planning projects initiated by state and local government emerged, including the City of Brisbane Draft Town Planning Scheme (1944); the County of Cumberland Planning Scheme, encompassing the Greater Sydney region (1948); and the Melbourne Metropolitan Planning Scheme (1954). A close examination of these and other regional land use planning initiatives reveals no connection between them and the methodological approach of the environmental (landscape) resource mapping initiatives pioneered by CSIRO in the post-WWII era. There appears to have been limited detailed analysis of the biophysical aspects, especially the constraints and opportunities provided by those environmental attributes for future urban development. If the biophysical elements were considered, they were treated separately and were not seen as part of a holistic landscape unit.

Planning for the 1960s and 1970s

Planning for the next two decades of regional-scale land use endeavours witnessed the intensification of resource mapping and planning from time to time by various state and national governments. Of particular note at the national level were the Whitlam Labor Government's Growth Centres and New Cities planning initiatives. These bold urban and regional planning forays into hitherto state-dominated affairs saw joint federal–state nominations of growth centres, accompanied by a promise of federal funding subsequent to the preparation of land use plans that met Australian and state government objectives. Some of the major growth centres included Gosford–Wyong (NSW), Bathurst–Orange (NSW), Albury–Wodonga (NSW–Victoria), Monarto (SA), Darwin New Town (NT) and the Moreton Region (QLD). In subsequent years, the Australian Government's interest in metropolitan and regional land use planning waxed and waned, with inertia being the dominant characteristic. Meanwhile, most state governments completed various metropolitan- and regional-scale strategic plans during this period of modest growth; an exception was the Brisbane City Council, the country's largest single metropolitan local authority, which completed its own plan.

Lost Opportunities

There were numerous attempts to plan at the regional- and metropolitan-scale throughout the post-WWII period, and these continued into the 1960s and 1970s. These initiatives, which involved all levels of government, were of a strategic planning nature that sought to satisfy a number of similar objectives, including the identification of areas for future growth and the future form of cities and regions, and the assignment of broad land use allocations across the landscapes in question. It could be assumed that these broad strategic planning objectives were informed by a detailed understanding of the biophysical attributes of the respective planning areas. Further, it could be assumed that these physical planning endeavours might have accessed existing environmental (landscape) resource mapping initiatives of the type undertaken by CSIRO. Yet this did not always occur, and it is puzzling as to why planners did not employ similar methodological approaches to inform their planning process. The question of an assumed association between early land resource studies and regional and metropolitan land use planning initiatives is even more relevant when it is noted that the former were more than just natural resource inventories—in many cases, they involved assessments of the physical attributes of the landscape to accommodate various forms of urban and infrastructure development. For example, the CSIRO division of soil mechanics' PUCE Terrain Evaluation for Engineering Purposes series covered several metropolitan regions, including Brisbane, Sydney, Melbourne and Perth. However, there is little evidence that the various land use planning initiatives were informed by these resource mapping initiatives, or hybrids based on their methodology. If they were, then other—non-environmental—considerations overrode the environmental (biophysical) values in the final land use planning process; this is what occurs in political decision-making processes.

The classic case in point is the tragic loss of the highly valuable krasnozems and red earth soils that formerly dominated elevated locations on the fringe of Brisbane city. These areas supported productive market gardens; however, they were also sought after for 'greenfield' land development due to their elevated locations, proximity to Brisbane and location within the commuting zone of increasingly affluent city workers. The ability of the land development industry to outbid farming interest in the purchase of these areas was also a factor. Hence, in the planning processes for these areas, economic values and the interests of land developers took precedence over biophysical–landscape values and agricultural interests, and these productive fruit- and vegetable-growing areas were lost forever. This situation has been repeated in many other sprawling urban centres. While the question of environmental (landscape) values is complex (particularly their place in the even more complex planning and policy decision-making process), it is contended that consideration of the biophysical attributes of the landscape should have played a greater role in the planning process.

Putting aside the political processes, the apparent lack of a nexus between land use planning and environmental resource studies raises a number of questions:

1. Were the planners of this era aware of these various examples of technical environmental–landscape data upon which sound land use planning decisions should have been made?
2. If they were, why didn't they access such input or commission such studies?
3. Were land use planners capable of interpreting this data and incorporating it into their land use planning processes?

Challenges of Traditional Land Use Planning

In reviews of attempts to address past environmental and development challenges through planning, several shortcomings of traditional land use planning have been noted. Although it can embrace a broad range of planning endeavours, for the most part, traditional land use planning education and practice has been (and continues to be) almost exclusively associated with the design professions, such as architecture (i.e. it does not have strong environmental–biophysical underpinnings). Critics of the theory and practice of traditional forms of planning have noted the following shortcomings:

- physical and design bias (Taylor, 1998)
- normative approach that overemphasises utopian ideals; conservative concern for aesthetics; promotes a 'technicalist' view of planning (Taylor, 1998)
- singular urban and economic efficiency concentration (Herring, 1999) and lack of rural focus (Laut & Taplin, 1988)

- underpinned by laws, regulations and guidelines that were developed to protect society from human error, and for health, safety and welfare reasons, rather than environmental sustainability (Forman, 1995)
- lack of a suitable philosophical perspective to address emergent environmental management and ecological issues (Conacher & Conacher, 2000)
- lack of adequate science input (Johnson, Swanson, Herring & Greene, 1999).

Consistent with these shortcomings, one critic observed that:

> Strategic planning needs to be based on a better understanding of local and regional physical, social and economic environments in Australia's coastal zone, and this information must provide the basis for planning and managing of development. (Graham, cited in Resource Assessment Commission, 1993, p. 96)

On the question of whether planners were 'capable of interpreting that data and incorporating it into the land use planning process', it was noted that there was a 'lack of information and poor communication channels between scientists and managers' (House of Representatives Standing Committee on Environment Recreation and the Arts, 1991, p. 42). Opdam, Foppen and Vos (2002) also acknowledged the lack of a nexus. However, they saw the challenge from the opposite perspective, pointing out that:

> Most empirical process studies are of no use to landscape management as long as we fail to transfer the information to the level of problem solving ... [We must ensure] that this gap between process studies and spatial planning is bridged ... [The] lack of a mechanistic basis for a holistic landscape ecology and, consequently, for spatial planning, is because many empirical and theoretical ecological studies fail to transfer their results in the context of landscape pattern. (pp. 767–68)

As landscape managers, land use planners need to appreciate the value of biophysical information in their planning decisions; they need to be aware of the existence of this data; they need to be able to commission similar natural resource (landscape) studies; and, importantly, they need to be able to interpret biophysical scientific data for planning purposes.

An Alternative Planning Paradigm (1970s and 1980s)

While not an entirely new paradigm, the landscape (ecological) planning approach was given renewed emphasis from the early 1970s with the publication of Ian McHarg's (1969) seminal work *Design with Nature*. McHarg wrote:

> We need nature as much in the city as in the countryside. In order to endure we must maintain the bounty of that great cornucopia which is our inheritance. It is clear that we must look deep to the values which we hold … these must be transformed … to do this [we] must design with nature. (p. 5)

McHarg's advocacy coincided with the Whitlam Government's Growth Centres planning initiative that, as noted, rarely connected the land use planning function with scientific landscape assessments that were capable of informing the planning process. The same was true of background studies used to support the raft of other (largely state-initiated) regional- and metropolitan-scale planning studies of this era. Although a number of state-based systems did adopt, or modify, CSIRO's land resource methodology, these were mostly agencies that had strict conservation planning functions for state lands only (e.g. the Victorian Land Conservation Council [now Victorian Environmental Assessment Council]) or had no mainstream land use planning function in relation to land development (e.g. Department of Primary Industries). Reflecting on lost opportunities, when *Design with Nature* was republished in 1992, McHarg observed that:

> In 1969, while many people accepted the proposition—Design with Nature—there was no legislation empowering or requiring ecological planning … now the situation is vastly different and it is the new legislation which provides this [new] book with an enlarged purpose … the power to employ ecological planning from national to local scales has accumulated slowly. Serious omissions remain, notably the fragmentation of environmental sciences and the plethora of responsible institutions, but there are now innumerable opportunities to employ the method. (1992, p. vi)

Contemporary Planning (1990s–present)

McHarg's observation sums up the current regional planning situation in Queensland. The South East Queensland (SEQ) region, which contains the state capital of Brisbane city, has experienced phenomenal population growth by Australian standards over the last decade and is expected to remain the fastest growing metropolitan region in the country for at least the next decade. Strategic regional planning for SEQ has been completed under the auspices of Queensland's *Sustainable Planning Act 2009* (as amended), which, like almost all state and territory planning legislation, is underpinned by the principles of ecologically sustainable development. Sustainable development aims to provide for the social and economic needs of society, while protecting environmental resources and values for the future (World Commission on Environment and Development, 1987).

Featuring 'values-led' planning, this paradigm advocates for an evidence-based approach that leads to science informing both planning and the values upon which planning is predicated. In an environmental context, values have been defined as 'direct and indirect qualities of natural systems that are important to the evaluator' (Satterfield, 2001, p. 332). Contemporary planning processes and practices can facilitate a values-led approach, whereby the vision and aspirational statements that guide and direct the plan attest to the collective values of those communities engaged through the planning process.

Low Choy (2015) showed how the values of communities that seek a high degree of liveability are associated with planning processes that ensure that key landscape attributes (i.e. attributes that have a positive influence on the achievement of the aspirational vision of liveability) are protected. This relationship between community (environmental) values and the achievement of community aspirations of liveability through the protection of key landscape attributes is illustrated graphically in Figure 5.1.

Values-led planning acknowledges that planning operates within a political and often highly contested context, which invariably involves contestations about the different values that sections of a community hold. Planning should be led by contemporary values, and these should be elicited through appropriate community engagement processes that are informed by robust evidence, including the best available science. In this way, regional-scale strategic land use planning can incorporate landscape values that have been informed by science that is consistent with biophysical (landscape) data, as exemplified by the environmental (landscape) reports of the post-WWII resource mapping initiatives (see Queensland Government, 2009).

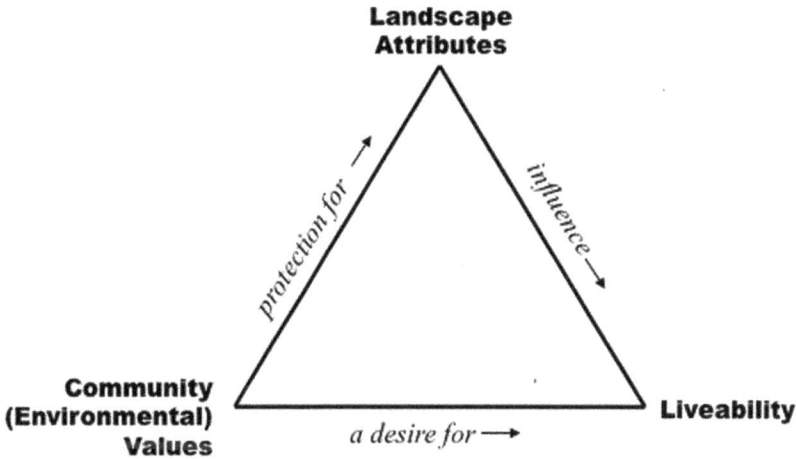

Figure 5.1: Relationship between landscapes, values and a community goal of liveability.
Source: Low Choy (2015).

Recommendations

Conventional land use planners are not trained as scientists; they do not come from a scientific background and many have limited experience in dealing with scientists. Further, scientists have not traditionally collaborated with land use planners. Science-informed planning paradigms were very rarely advanced in the past; however, there have been attempts to seek robust and sustainable outcomes in recent years, as exemplified by numerous contemporary metropolitan planning schemes. Enhancing the nexus between planning and science needs to be addressed, in the first instance, through education. Both disciplines need to be modified and this will take time. In the meantime, the following strategic principles for integrating science and planning will go some way towards strengthening the nexus:

- incorporate scientific analysis and review throughout the entire planning process
- integrate scientists and their work into the planning process from the beginning

- respect scientific independence: scientists must remain objective
- planners must clarify the specific type and scale of scientific expertise and information required for the planning initiative in question
- scientists must provide their work in 'user-friendly' formats
- planners must put forward credible technical solutions, based on best available science, to have the greatest chance in competing against other interests in the political decision-making process.

These initiatives will succeed if they are implemented under the auspices of overarching planning reform at national and state levels. At the national scale, an opportunity for the future has been advanced by the Wentworth Group of Concerned Scientists (2014), an independent group of Australian scientists, economists and business people concerned with advancing solutions to secure the long-term health of Australia's land, water and biodiversity. In November 2014, the Wentworth Group released their 'Blueprint for a Healthy Environment and a Productive Economy'. In this national-level response to contemporary land, water and biodiversity challenges, the Wentworth Group advanced five transformative, long-term economic and institutional reforms—two of which are relevant to this chapter:

1. Fix land and water use planning:

We must put in place regional scale land and water use plans that address the cumulative impacts of development on the environment and the long-term costs to the economy.

…

4. Regionalise management:

We must embed and give prominence to natural resource management at the regional scale to reconnect people to the land, so that investment decisions are underpinned by an understanding of how landscapes function. (p. 2)

Ideally, future land use planners will better appreciate the valuable role that science (especially biophysical science) can have in informing the planning process, and they will be capable of accessing and utilising scientific input into their planning endeavours. This, in turn, will lead to more robust and politically argued outcomes. The future should also witness a reciprocal understanding between planners and scientists, in which scientists freely engage with planners, communicating their science in ways that planners can use for planning purposes. Holistic and environmentally based planning, with a strong nexus between land use planning and environmental sciences, will require a 'marriage' between planners and scientists. As the past has shown with respect to traditional land use planners, this necessitates a concerted effort to understanding that 'you don't know what you don't know'.

References

Christian, C. S. & Stewart, G. A. (1952). *A survey of the Katherine–Darwin region, 1946.* Land Research Series: No 1. Canberra, ACT: CSIRO Publishing.

Conacher, A. & Conacher, J. (2000). *Environmental planning and management in Australia.* Melbourne, VIC: Oxford University Press.

CSIRO Publishing. (n.d.). *CSIRO land research surveys: Mapping land resources in Australia and Papua New Guinea.* Retrieved from www.publish.csiro.au/nid/289.htm

Freestone, R. (2012). *Post-war reconstruction and planning promotion in 1940s Australia.* Paper presented at the 15th International Planning History Society Conference, Sao Paulo, Brazil, July 2012.

Forman, R. T. T. (1995). *Land mosaics—the ecology of landscapes and regions.* Cambridge, UK: Cambridge University Press.

Grant, K. (1971). Terrain evaluation for engineering purposes. In *Proceedings of symposium on terrain evaluation for highway engineering, special report no 6* (pp. 81–107). Melbourne, VIC, Australian Road Research Board.

Herring, M. (1999). Introduction. In K. N. Johnson, F. J. Swanson, M. Herring & S. Greene (Eds.), *Bioregional assessments—science at the crossroads of management and policy.* Washington DC: Island Press. doi.org/10.1002/j.2164-490x.1999.tb00063.x

House of Representatives Standing Committee on Environment Recreation and the Arts. (1991). *The injured coastline: Protection of the coastal environment.* Canberra, ACT: Australian Government Publishing Service.

Johnson, K. N., Swanson, F. J., Herring, M. & Greene, S. (Eds.). (1999). *Bioregional assessments—science at the crossroads of management and policy.* Washington, DC: Island Press.

Laut, P. & Taplin, B. J. (1988). *Catchment management in Australia in the 1980s.* Canberra, ACT: Division of Water Resources, CSIRO.

Low Choy, D. C. (2015). *Managing ecosystem services in the peri-urban landscape: An emergent paradox.* Paper presented at the Second International Conference on Agriculture in an Urbanizing Society, Italy, September 2015.

McHarg, I. L. (1969). *Design with nature.* New York, NY: Doubleday/ Natural History Press.

McHarg, I. L. (1992). *Design with nature.* New York, NY: John Wiley.

Opdam, P., Foppen, R. & Vos, C. (2002). Bridging the gap between ecology and spatial planning in landscape ecology. *Landscape Ecology 16,* 767–79. doi.org/10.1023/A:1014475908949

Queensland Government. (2009). *South East Queensland regional plan 2009–2031.* Brisbane, QLD: Department of Infrastructure and Planning.

Resource Assessment Commission. (1993). *Resource Assessment Commission coastal zone inquiry: Final report.* Canberra, ACT: Australian Government Publishing Service.

Satterfield, T. (2001). In search of value literacy: Suggestions for the elicitation of environmental values. *Environmental Values 10*(3), 331–59. doi.org/10.3197/096327101129340868

Taylor, N. (1998). *Urban planning theory since 1945.* London, UK: Sage Publications.

Wentworth Group of Concerned Scientists. (2014). *Blueprint for a healthy environment and a productive economy.* Retrieved from wentworth group.org/wp-content/uploads/2014/11/Blueprint-for-a-Healthy-Environment-and-a-Productive-Economy-November-2014.pdf

World Commission on Environment and Development. (1987). *Our common future.* New York, NY: Oxford University Press.

6

Responding to Land Use Pressures: A State and Territory Perspective

Richard W. Hicks

Key Points

- Under the Australian Constitution, each state and territory has responsibility for land use policy, planning and management.
- Each state has varying legislative drivers and land use pressures that present a range of challenges, including:
  - biodiversity legislative reforms, which impact on planning controls of land use changes by restricting land use intensification, particularly as it relates to native vegetation management
  - controlling the expansion of urban areas into intensive high-value agricultural lands
  - mitigating the impacts of mining developments on surrounding land uses and rehabilitating mined areas to provide functioning ecosystems that can support viable land uses
  - protecting waterways to ensure clean water supply and sustainability of the aquatic and marine environments.

- Over the past 16 years, the state jurisdictions and relevant Australian government agencies have worked together to develop a comprehensive and consistent land use framework under the auspices of the Australian Collaborative Land Use and Management Program (ACLUMP).

- A consistent national framework for land use classification and mapping has strengthened the states' commitment to preparing state-wide land use information, resulting in the first national compilation of catchment-scale land use data in 2008.

- Most states are remapping or upgrading the precision of their original mapping to reflect this national framework. Under this national collaborative framework, all jurisdictions now share land use information, enabling a national report on land use change and trends to be prepared.

- Emerging trends arising from advances in technology, especially in remote sensing and computing technologies, are enabling improved monitoring and reporting of land use changes and trends.

The information presented in this chapter is largely based on information compiled from agency representatives involved in ACLUMP. ACLUMP is a partnership between Australian, state and territory government land management agencies, as well as relevant research organisations. This program promotes the development of nationally consistent land use and land management practices information for Australia.

Responding to Land Use Pressures

Under the Australian Constitution, the states and territories have responsibility for land use and management across a range of public and private land tenures. Changes to land use policy and planning can have a major bearing on the natural resource condition of land, water, air, soil and native vegetation. Changes in resource condition often have a long lag time—years or decades can pass before the ecological effects of changes in land management regimes and practices are observed (Thackway & Freudenberger, 2016). Public–private responses to issues of declining resource condition are many and varied. They include targeted policy and planning responses to issues of salinity, poor water quality and the maintenance of biodiversity (e.g. Murray–Darling Basin planning and numerous land care strategies implemented by Australian and state governments over the past 30 years). This involves investigation of land use, land cover and land management, and developing trade-offs between competing use options and land management regimes (Lesslie, Barson & Randall, 2008).

Before the 1980s, information that enabled the development of land use policy and planning instruments was generally derived from land system and soil maps (e.g. Department of Agriculture, 1985; Weston, Harbison, Leslie, Rosenthal & Mayer, 1981). Since then, public agencies have produced maps of land use types and their extent by combining several data types: land use patterns derived from remotely sensed imagery, biophysical information, social and economic information, and ground-based surveys (e.g. Department of Conservation and Land Management, 1992; Department of Water Resources, 1989). These land use related datasets are employed by local, state and Australian government agencies to identify opportunities for appropriate intensification of land uses and improvements to land management regimes and practices. Evaluations of the outcomes of changing land use in combination with better management approaches have led to widespread and significant declines in dryland salinity and associated problems, decreased rates of wind and water erosion, and improved protection and management of native vegetation.

Development of ACLUMP

Before 1999, the availability of detailed mapping of land use in Australia was limited and uncoordinated across the various jurisdictions. Australian and state government agencies independently produced land use mapping at a range of scales using a variety of cartographic methods and classification systems.

ACLUMP was established in 2000 in response to increasingly complicated and complex land use policy and planning issues facing all Australian land management jurisdictions, including food security, vegetation and carbon management, biosecurity, climate change, sustainable agriculture and water management. At the same time, Australia started systematic reporting in respect to forest cover and national carbon accounting as a result of the Kyoto and United Nations reporting requirements. Each of these issues called for coordinated and cooperative approaches to developing a responsive national land use infrastructure. Key elements of the infrastructure included mapping, coordination and standards, communication and dissemination, and analysis and reporting. These elements are described in more detail below.

ACLUMP is overseen by a national committee representing Australian and state government agencies. The program promotes the development and use of a nationally consistent set of land use and land management practices and reporting codes throughout Australia. The principles of classification, decision rules for mapping and classification evolutions are continuously updated—the current version is 8.0 (Australian Bureau of Agricultural and Resource Economics and Sciences [ABARES], 2016).

In 1999, the National Land and Water Resources Audit, Department of Agriculture, Fisheries and Forestry (DAFF), Bureau of Rural Sciences (BRS), Murray–Darling Basin Commission and state agency partners commenced a collaborative national land use mapping initiative that led to the development of ACLUMP. DAFF accepted leadership for the national coordination of land use information in 2000, and BRS, later ABARES, took on responsibility for the development of ACLUMP. Mapping products are now in strong demand for a range of land management purposes, and there is widespread adoption of agreed standards. The importance of ACLUMP and the information it provides is widely recognised. For example, a recent report to the Australian Farm Institute (Budge et al., 2012) recommended that ACLUMP be strengthened and further supported.

ACLUMP's partners are well advanced in meeting catchment-scale (nominally 1:50,000 or 1:25,000 in intensively used areas) land use mapping goals. These are:

1. Comprehensiveness: continental mapping coverage at the catchment scale (excluding parts of the ACT) was completed in 2008. Since then, most jurisdictions have remapped areas. The last release of catchment-scale data compiled for Australia was in March 2015 (see Figure 6.1).
2. Accuracy: with some limited exceptions, all land use products have an overall (or total) accuracy of greater than 80 per cent. For jurisdictions that do not directly map using the Australian Land Use and Management (ALUM) classification (e.g. NSW and VIC), conversion tables have been developed to ensure that land use can be mapped consistently across Australia.

3. Currency: ACLUMP partners have determined priorities for updating land use mapping. The mapping for many of the intensive agricultural areas is greater than five years old, while those for pastoral areas in Queensland and Western Australia are greater than 10 years old (see Figure 6.2). ACLUMP partners are addressing data currency through their respective state-based programs in accordance with the priorities identified, and in response to key policy drivers in each jurisdiction.

4. Scale: one of the major strengths of catchment-scale land use mapping is the high resolution at which it is produced. Most intensive agricultural areas are mapped at either 1:25,000 or 1:50,000, with a minimum mapping area of 2 hectares.

The primary challenges for land use mapping are maintaining accuracy and currency. However, given improvements in satellite imagery, especially the frequency of updates, as well as access to platforms such as Google Earth, these challenges can easily be addressed, as discussed below. The major focus is on measures that enable the use of ancillary data and land use change assessments, which contribute to improved accuracy and currency of catchment land use.

ACLUMP is well placed to strategically respond to land use and management challenges as they arise; this is due to the coordinated national network of state partners involved in the program. Recent increases in the extent of irrigated agriculture, vineyards and cereal cropping in some regions have raised the need for improved monitoring and reporting of land use changes and trends. A perennial problem for local and regional policy and planning are the land use issues associated with peri-urban and coastal areas, whereby land uses can change rapidly in response to the pressures of urbanisation (Lesslie et al., 2008).

Several states and territories are facing pressures of declining water availability, loss of soil carbon associated with broadscale agriculture and loss of biodiversity. Intensification of land uses is leading to demands from users for finer-scale and more current land use datasets. Map sheets for regions undergoing substantial change need to be updated regularly; this is increasingly being addressed by incorporating changes detected via remote sensing, or other spatially explicit data, into the existing land use datasets, and updating the metadata and validating the new land use map.

The regular updating of data and information enables states to monitor and report changes and trends in land use; most states have undertaken to update their initial mapping, either via a second round of mapping or by using recent imagery.

Predictive modelling of natural resource management issues at the catchment level requires information about land management regimes and practices. In 2004, agreement was reached among relevant government departments, industry groups and scientific organisations on the need to develop a national categorisation and information system for land management practices (ACLUMP, 2010a, 2010b); however, the Land Use Management Information System (LUMIS) has not been fully developed or realised. Instead, in 2010 (or 2011), ACLUMP shifted its focus to improving its understanding of (and consistency in) the monitoring of ground cover changes across Australian landscapes. This shift was driven by the availability of new remote-sensing datasets that measured and reported ground cover changes and trends at a range of spatial and temporal scales. Recognition of the general decline in land condition across the continent due to ground cover management practices—particularly with respect to soil health, wind and water erosion, but also the effects on agricultural productivity—was another factor.

Having established a national ground cover monitoring program, ACLUMP has recently refocused its efforts to updating and improving land use mapping, particularly in those regions that are long outdated or have experienced significant or rapid land use change. While these developments are welcome, it remains difficult to distinguish local variations in ground cover that are due to management inputs from those that are due to rainfall extent and timing. It is also difficult to remotely identify various land management practices and the rapid changes occurring in cropping practice, due to the ongoing development of precision agricultural systems. This demonstrates the need for further calibration and validation of ground-cover mapping products.

Benefits of Adopting and Implementing the ACLUMP Program

National Benefits

ACLUMP's overarching goal is to develop a forum to support an integrated land use and land management practices information system for Australia. Implementation of the nationally agreed ACLUMP approach has produced many benefits for the states and territories, such as:

- mapping: nationally consistent land use mapping for Australia at both national and catchment scales
- coordination and standards: agreed technical standards, including ALUM and LUMIS
- communication and dissemination: a national land use data directory and the maintenance of land use datasets on Australian and state government data repositories
- analysis and reporting: regional and national reporting of land use and land management practices, including change reporting and integrated assessments.

At the state level, land use information is collected to ACLUMP standards using the ALUM classification. This allows consistent reporting at the catchment scale, as well as at state, territory and national levels. Land use mapping technology and display at the desktop level may vary between jurisdictions; however, they all meet the agreed minimum standards collaboratively developed by ACLUMP's partners (Lesslie, Mewett & Walcott, 2011; Mewett et al., 2013).

The land use information derived by the various state programs is used to support urban and regional planning, biodiversity regulations, bushfire planning and biosecurity planning, among other uses (see Figure 6.1) (OEH [Office of Environment and Heritage], 2016d). Other state and territory responsibilities include food security, vegetation and carbon management, biosecurity, climate change adaptation, sustainable agriculture and water management.

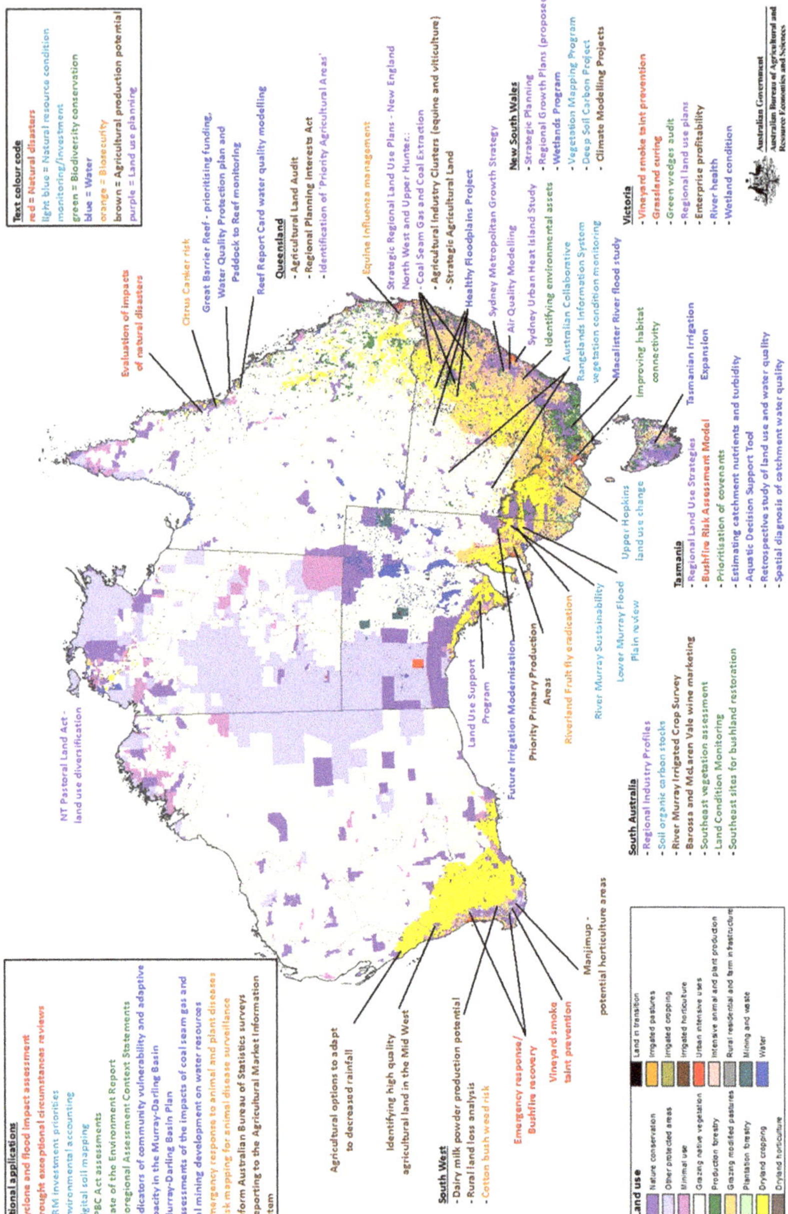

Figure 6.1: Applications of catchment-scale land use data.
Source: ABARES (2015, Figure 1).

In regard to collecting, monitoring and employing data and information relevant to land use policy and planning, some states rely on intensive manual interpretation of aerial and satellite imagery—such as MODIS (Moderate Resolution Imaging Spectroradiometer), Landsat, SPOT 5 and Sentinel 2—with appropriate field verification; with the exception of SPOT 5, generally every pass is collected. Other states rely on semi-automated imagery analysis to detect land use changes from the original land use mapping (OEH, 2016b). Utilising research undertaken by the Joint Remote Sensing Research Program provides the opportunity to use Landsat- and SPOT 5–based multi-temporal information to detect the probability of land use change through woody change detection and seasonal fractional cover analysis.

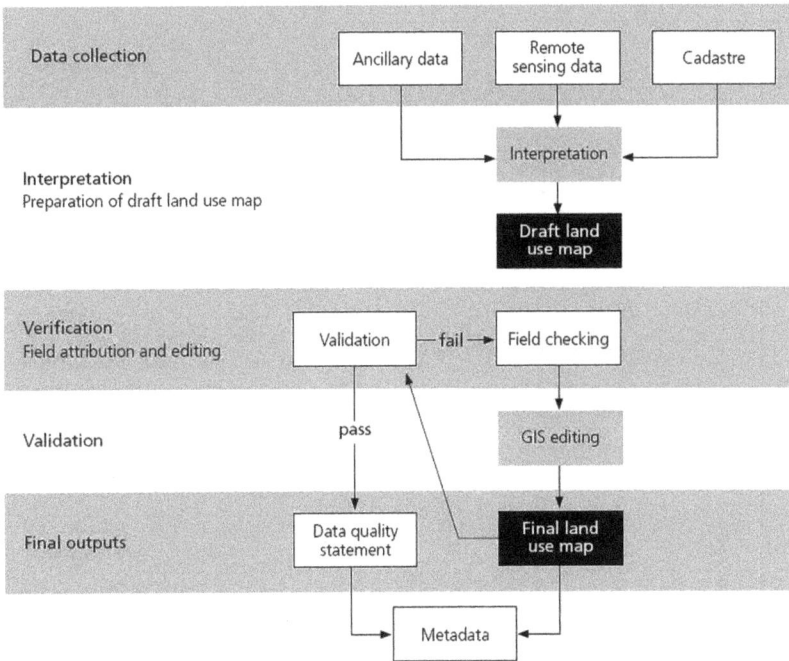

Figure 6.2: Land use mapping procedure.
Note: GIS = geographic information system.
Source: ABARES (2011, Figure 6).

Mapping techniques are rapidly adapting to new innovations, such as aerial and satellite imagery available through Web Map Services (WMS) and geographic information system (GIS) plugins. Examples include Planet Labs (satellite), Nearmap (aerial) and Google Earth (aerial and satellite). WMS provide imagery to consumers much more quickly than traditional imagery acquisition techniques. This expansion in data types and accessibility enables jurisdictions to more rapidly update their land use data holdings. To ensure consistency across jurisdictions in land use mapping products, an Australian land use classification and generic mapping procedure was developed (see Figure 6.2).

Intrastate and Territory Benefits

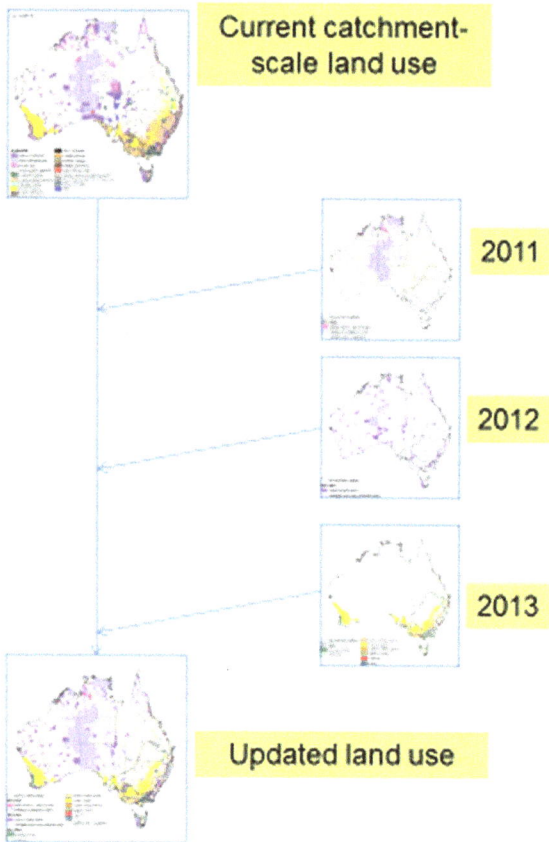

Figure 6.3: Updating land use mapping by incorporating changes detected by acquiring annual imagery.
Source: Randall, Mewett & Purcell (2015, based on Figure 3).

Each state has different priorities and available resources to undertake land use mapping. The national ACLUMP standards provide the framework for the different state-based land use mapping programs. Using the

ACLUMP framework for land use classification and mapping, the larger states can collect annual (or every pass of available) imagery. This access to regularly available new data is driving an 'updating by exception' approach to land use and land cover mapping (see Figure 6.3), whereby states update the underlying datasets by identifying areas of change using satellite and recent aerial imagery. This approach, which does away with the need for new mapping programs to update land use, is made possible with the help of modern, high-capacity remote-sensing computing systems. It assists with updating land use mapping in a resource-constrained environment by utilising authoritative ancillary datasets for specific land uses, and has enabled both larger and smaller states and territories to maintain their land use mapping programs (see ABARES, 2015).

Interstate Benefits

The adoption of a consistent system for classification and mapping provides an objective, robust and defensible framework for national and state land management initiatives. This has been brought about by:

- using a common set of terms and descriptions
- sharing common database attributes and algorithms, enabling increasing levels of interoperability, and efficiency gains in mapping approaches between state and national programs
- improved understandings of common techniques and landscape impacts associated with land management practice across multiple jurisdictions (e.g. cultivation practices such as minimum tillage and direct drilling)
- awareness that patterns in land cover can relate to management practice and land use.

Land Use Case Studies

The need for more comprehensive information on land use changes, especially in the larger states, is driven by issues associated with the protection of native vegetation and rapid urban expansion into surrounding agricultural areas. The following two examples, which highlight rapid change in land use over the past 10 years, show how land use and land management information has been deployed by government at local, regional and state levels.

Example 1: Moree Floodplains, NSW

The primary driver for the Moree floodplains analysis was the NSW annual woody change detection program (part of the Statewide Landcover and Tree Study) that used annually acquired SPOT 5 satellite imagery. Cropping activity in the Moree floodplains, both dryland and irrigated, increased from 854,000 hectares in 2003 to 939,100 hectares in 2013—nearly a 10 per cent increase in cropping activity over 10 years.

Figures 6.4a, 6.4b and 6.5 show the expansion in cropping on the Moree floodplains over 10 years. The predominant land use change was the conversion of grazing lands, including areas that were covered by woody vegetation, into dryland and irrigated cropping. These insights were made possible by developing a time-series land use change dataset. Between 2004 and 2012, 6,485 hectares of woody vegetation change (loss) was attributed to agricultural activity for this study area. The extent of woody vegetation change detected for the study area is shown in Figures 6.4a and 6.4b.

For the areas identified as grazing intensification (Figure 6.5), 12,100 hectares were marked as grazing of native vegetation (land use c. 2003), whereby the woody component had been removed. It is likely that these areas are in a transitional phase, and that they will become areas of cropping activity in the future (see Table 6.1). These areas would be the focus for future land use updates. This type of information has informed policy directions being enacted by the NSW Government (e.g. recent biodiversity conservation legislation changes), especially with regard to the implementation of legislative controls over the loss of native vegetation and conversions of grazing land uses to cropping.

This study of changes in land use has been used to assess the impacts of cultivation expansion on natural resources: soil, water and extent of native vegetation. It has also been used to assess the impacts on biodiversity, groundwater and flood-dependent ecosystems in over-cleared landscapes, based on the NSW Mitchell Landscapes version 3 (OEH, 2011). In addition, it has been used in the NSW Healthy Floodplains Project—Floodplain Management Plans (Department of Primary Industries, 2011) and state vegetation-type mapping and landscape modelling (OEH, 2016a). It was a precursor scoping study for the NSW Biodiversity Act Map, as part of the NSW Biodiversity Legislation Review.

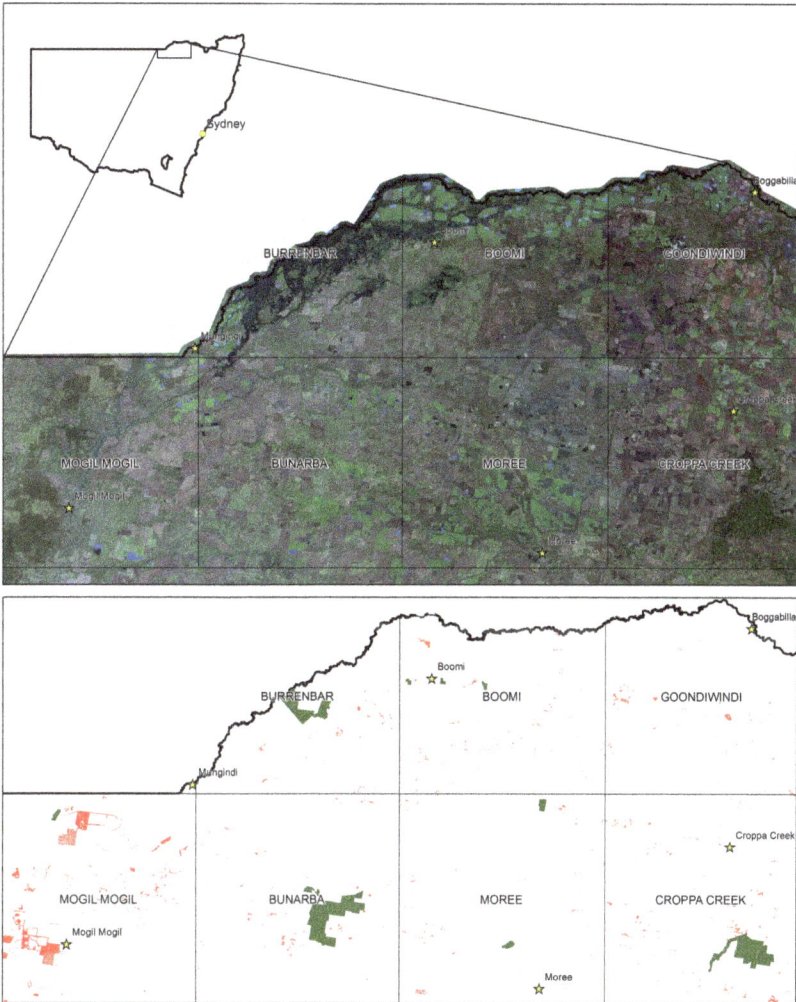

NSW SLATS Woody Change 2004 -2012

■ Agricultural Woody Change
■ Reserve

Figures 6.4a and 6.4b: Extent of land use changes on the Moree floodplains (1:100,000 topographic map sheets illustrated), showing the loss of woody vegetation, 2004–12, New South Wales Statewide Landcover and Tree Study Program.

Source: OEH (2016b), used with permission.

Table 6.1: Cultivation expansion to 2013 on the Moree floodplain and the loss of ALUM land use types based on existing (2003) land use information for New South Wales.

Loss of:	Cultivation expansion and the loss of land use types (NSW land use, 2003)								
	Natural Environment (ALUM 1)	Marsh/ Wetland (6.5.0)	Rivers (6.3.0)	Lake (6.1.0)	Grazing Native Vegetation (ALUM 2)	Grazing Modified Pastures (3.2.0)	Reservoir Dams Channels Aqueducts (6.4.0)	Land in Transition (3.6.0)	Intensive Uses (ALUM 5)
Area (ha)	1,070	610	110	210	46,990	31,730	350	580	30

Note: ALUM = Australian Land Use and Management.

Natural Environment ALUM primary class one changes were associated with 1.3.3 – areas mapped as native residual cover, where the overstorey and understorey were mapped as being relatively intact.

Source: OEH (2016b).

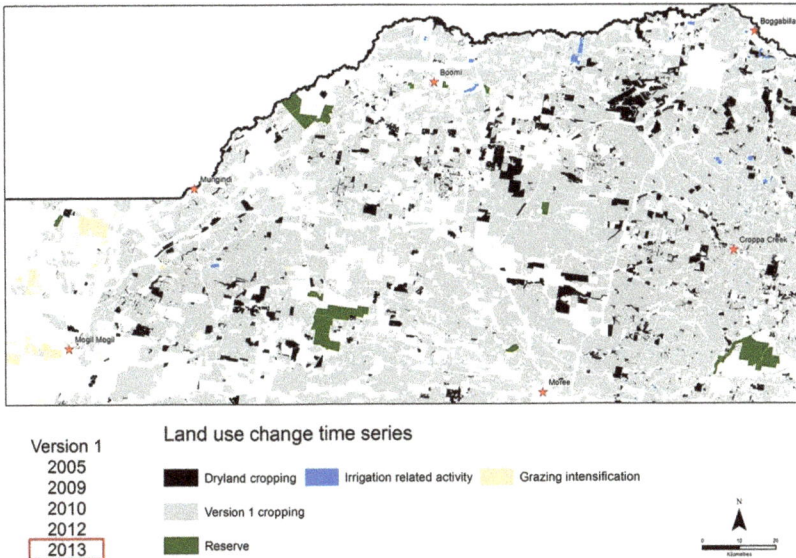

Figure 6.5: Extent of land use change on the Moree floodplains, 2003–13.

Note: The light grey shows the cropping extent in 2003. The expansion of dryland cropping is shown in black and irrigated cropping in blue.

Source: OEH (2016b).

Example 2: Maitland Local Government Area

Maitland has undergone a rapid increase in population in recent years; indeed, it has experienced the largest population growth among inland centres in NSW. A service centre for Newcastle and the Hunter Valley, and located within easy commuting distance of Newcastle and the Hunter Valley coalmines, its commercial areas have expanded to service the mining industry and to provide household retail outlets (see Figures 6.6, 6.7a and 6.7b).

Updated land use information for the Maitland local government area and Hunter Valley was provided to the NSW Departments of Planning and Environment (DPE) and Primary Industries for the Upper Hunter Strategic Regional Land Use Plan (UHSRLUP) (DPE, 2012) (see Table 6.2). The information was used to assist in identifying land use activities present in the Upper Hunter Valley, including the extent of mining operations and critical industries clusters (agriculture) defined in the UHSRLUP. Land use information is now critical for future regional planning in Australia and is actively considered in the planning processes used by state planning agencies.

Figure 6.6: Maitland local government area.
Source: OEH (2016b).

Table 6.2: Changes in land use for Maitland local government area at the ALUM primary classification level, based on existing (2003) NSW land use information and mapped changes (2013).

ALUM primary classes	2003 (%)	2013 (%)	% Difference
1. Conservation and natural environments	5.0	4.7	−0.3
2. Production from relatively natural environments	0	0	0
3. Production from dryland agriculture and plantations	70.7	64.4	−5.6
4. Production from irrigated agriculture and plantations	4.9	2.8	−2.1
5. Intensive uses	13.5	21.5	8.1
6. Water	5.8	6.6	0.8

Note: ALUM = Australian Land Use and Management

Loss of ALUM primary class 1 is attributed to the loss of forested areas 1.3.3−residual native cover (NSW land use, 2003).

Source: OEH (2011).

Figures 6.7a and 6.7b: Decline in dryland and irrigated agriculture and increase in intensive uses. Residential, commercial and light industrial expansion around Rutherford, north-western edge of Maitland developed area.

Source: OEH (2016b).

Conclusions

The need for ongoing improvements to the datasets (and to abilities to correlate this information with other information) will only increase over time—for example, due to the growing demand for land use information to support biosecurity planning and emergency responses. In addition, the changes in land use associated with a changing climate are becoming apparent, with major shifts in cropping practices occurring in southern NSW and Victoria due to the decline in winter rainfall.

Australia's land stock, particularly in the southern half of the continent, is undergoing steady intensification of use as the population grows. This pressure requires an ongoing commitment by governments to provide quality information to support the required planning and development activities (Budge et al., 2012). In the northern half of Australia, considerable effort is being put into capability–suitability assessments to identify possible areas for intensification of agricultural output. Predictive modelling is now used as part of the planning process and it requires data on current land use and land management at scales appropriate to the problems being addressed. Owing to the open data policies now in place, maps of current land use and alternative land use scenarios also help communities to participate in proposed development planning.

Future advancements in land use mapping will come from improvements in satellite monitoring technology, such as improved resolution and repeat cycle times and improvements to automated mapping approaches (e.g. machine learning for change recognition and image segmentation technology enabling efficiencies, and improvements in the accuracy and objectivity of the state-based and national programs).

Land use mapping at catchment scale is now generally available in the more intensively used areas of Australia (i.e. non-arid zones) apart from Victoria, which is covered by 1:100,000-scale mapping. Nationally agreed methods, developed under ACLUMP, have provided for cost-effective production, making best use of pre-existing land use information contained in sources such as cadastre (i.e. property-boundary information), public land and valuation databases, and land cover mapping programs.

Coordinated land use information relevant for policy and planning is critical to support state and national issues, such as biosecurity, as it provides underpinning land use location data for specialised industry activities:

for example, detailed mapping of the location of banana plantations across Queensland, NSW and Western Australia as part of the response to the Panama Tropical Race 4 disease outbreak. State agencies using ACLUMP to map those regions where there is a focus on horticultural commodities and intensive animal industries in collaboration with state biosecurity agencies and industry are another example.

Maintaining up-to-date and appropriately detailed information on changes in land use type and extent is an ongoing challenge for the states and ACLUMP. Such information is critical to informing debates around agriculture and food security, forestry, water, mining (including coal seam gas extraction), climate change mitigation and adaptation, population growth, urban expansion, biodiversity protection and landscape aesthetics. However, a better understanding of agricultural land use change is required for informed analysis of the future of agricultural production and land management in Australia.

References

ABARES (Australian Bureau of Agricultural and Resource Economics and Sciences). (2011). *Guidelines for land use mapping in Australia: Principles, procedures and definitions* (4th ed.). Canberra, ACT: Australian Bureau of Agricultural and Resource Economics and Sciences. Retrieved from www.agriculture.gov.au/abares/publications/display?url=http://143.188.17.20/anrdl/DAFFService/display.php?fid=pe_abares99001806.xml

ABARES. (2015). *Addendum to the guidelines for land use mapping in Australia: Principles, procedures and definitions* (4th ed.). Canberra, ACT: Australian Bureau of Agricultural and Resource Economics and Sciences. Retrieved from agriculture.gov.au/abares/publications/display?url=http://143.188.17.20/anrdl/DAFFService/display.php?fid=pe_agluma9abll20150415_11a.xml

ABARES. (2016). *The Australian land use and management classification version 8*. Canberra, ACT: Australian Bureau of Agricultural and resource Economics and Sciences. Retrieved from data.daff.gov.au/anrdl/metadata_files/pe_alumc9aal20161017.xml

ACLUMP (Australian Collaborative Land Use and Management Program). (2010a). *Land use and land management information for Australia: Workplan of the Australian Collaborative Land Use and Management Program*. Canberra, ACT: Australian Bureau of Agricultural and Resource Economics and Sciences, Canberra. Retrieved from www.agriculture.gov.au/abares/publications/display?url=http://143.188.17.20/anrdl/DAFFService/display.php?fid=pe_abares99001769.xml

ACLUMP. (2010b). *Status of land management practices activities of the Australian Collaborative Land Use and Management Program*. Canberra, ACT: Australian Bureau of Agricultural and Resource Economics and Sciences, Canberra. Retrieved from www.agriculture.gov.au/abares/publications/display?url=http://143.188.17.20/anrdl/DAFFService/display.php?fid=pe_abares99001770.xml

Budge, T., Butt, A., Chesterfield, M., Kennedy, M., Buxton, M. & Tremain, D. (2012). *Does Australia need a national policy to preserve agricultural land?* Surrey Hills, NSW: Australian Farm Institute.

Department of Agriculture. (1985). *Pastoral potential in the Kimberley region, Western Australia: 1:500,000 scale maps and notes* (Compiled from Land Systems Land Research Reports Nos 4, 9 and 28). Perth, WA: Rangeland Management Branch.

Department of Conservation and Land Management. (1992). *Land management proposals for the integrated treatment and prevention of land degradation—catchment map series*. Sydney, NSW: National Soil Conservation Program and Department of Conservation and Land Management.

DPE (Department of Planning and Environment). (2012). *Strategic regional land use plan: New England north-west*. Retrieved from www.planning.nsw.gov.au/~/media/Files/DPE/Plans-and-policies/strategic-regional-land-use-plan-new-england-north-west-2012-09.ashx

Department of Primary Industries. (2011). *Healthy floodplains*. Retrieved from www.water.nsw.gov.au/water-management/water-recovery/sustaining-the-basin/healthy-floodplains

Department of Water Resources. (1989). *Water Victoria: An environmental handbook*. Melbourne, VIC: Victorian Government Printing Office.

Lesslie, R. G., Barson, M. M. & Randall, L. A. (2008). Land use mapping. In N. J. McKenzie, M. J. Grundy, R. Webster & A. J. Ringrose-Voase (Eds.), *Guidelines for surveying soil and land resources* (2nd ed., pp. 141–55). Melbourne, VIC: CSIRO Publishing.

Lesslie, R. G., Mewett, J. & Walcott, J. (2011). Landscape in transition: Tracking land use change. *Science and Economic Insights, 2.2.* Retrieved from data.daff.gov.au/anrdl/metadata_files/pe_litlud9abll0790111a.xml

Mewett, J., Paplinska, J., Kelley, G., Lesslie, R. G., Pritchard, P. & Atyeo, C. (2013). *Towards national reporting on agricultural land use change in Australia* (Technical Report 13.06). Canberra, ACT: Australian Bureau of Agricultural and Resource Economics and Sciences. Retrieved from www.agriculture.gov.au/abares/publications/display?url=http://143.188.17.20/anrdl/DAFFService/display.php?fid=pb_nrlucd9ablm08320131011_11a.xml

OEH (Office of Environment and Heritage). (2011). *Floodplain policies and management.* Retrieved from www.environment.nsw.gov.au/floodplains/index.htm; www.environment.nsw.gov.au/resources/floodplains/130528Valleyplans.pdf

OEH. (2016a). *State vegetation type map.* Retrieved from www.environment.nsw.gov.au/vegetation/state-vegetation-type-map.htm

OEH. (2016b). *NSW woody change reports.* Retrieved from www.environment.nsw.gov.au/vegetation/reports.htm

OEH. (2016c). *Biodiversity conservation legislation—independent panel review.* Retrieved from www.environment.nsw.gov.au/biodiversitylegislation/review.htm

OEH. (2016d). *Biodiversity conservation legislation—native vegetation regulatory map development.* Retrieved from www.landmanagement.nsw.gov.au

Randall, L., Mewett, J. & Purcell, J. (2015). *Murray–Darling Basin land use project* (Report). Canberra, ACT: Department of Agriculture and Water Resources and Australian Bureau of Agricultural and Resource Economics and Sciences. Retrieved from data.daff.gov.au/anrdl/metadata_files/pc_mdblud9aal20150618_11a.xml

Thackway, R. & Freudenberger, D. (2016). Accounting for the drivers that degrade and restore landscape functions in Australia [Special Issue]. *Land 5*(4), 1–20. doi.org/10.3390/land5040040

Weston E. J., Harbison, J. K., Leslie J., Rosenthal, K. M. & Mayer, R. J. (1981). *Assessment of the agricultural and pastoral potential of Queensland.* Brisbane, QLD: Department of Primary Industries.

7

Balancing Land Use Trade-Offs: The Role of Wilderness in Nature Conservation

Jason Irving

Key Points

- Creating wilderness areas is often perceived in the community as an exercise in 'saving' areas, which can lead to conflict and a perception that wilderness protection is inconsistent with achieving other outcomes. Placing wilderness in its context as a core component of protected area systems and the several roles it plays in that context can help alleviate this. It can also reinforce the principles of wilderness that need to be maintained, such as the need for large-scale, intact areas.

- Pragmatism is inevitably necessary to achieve large-scale wilderness designations or active management of other tenured lands to maintain or enhance wilderness values.

- The evolution of the role of protected area systems, and the concept of wilderness, is not well understood or discussed, and neither are the challenges and policy adaptations for the Australian context of wilderness. Addressing this into the future is essential for maintaining the relevance of wilderness in conservation planning.

With over 20 years on the South Australian Wilderness Advisory Committee (WAC), which oversaw the creation of nearly 2 million hectares of wilderness protection areas, Rob Lesslie played a major role in promoting better understandings of the concept of wilderness, and applying it to protected area systems. The rapid growth in the wilderness estate between 2002 and 2012 led to land use trade-offs that were the subject of considerable debate. One of Rob's most powerful contributions was articulating how large wilderness areas could play a role in connecting natural areas and ecological processes across the landscape.

The South Australian *Wilderness Protection Act 1992* provides for the protection of wilderness and restoration of land to its pre–European contact condition. Areas defined as 'wilderness' under the Act share certain features: the land and its ecosystems must not have been affected (or must have been affected to only a minor extent) by modern technology, and must not have been seriously affected by exotic animals or plants or other exotic organisms. In assessing areas for wilderness protection, these criteria are applied using the conceptual framework that Rob pioneered for the Australian Government's National Wilderness Inventory. Four indicators of wilderness quality are tested: remoteness from access, remoteness from settlement, apparent naturalness and biophysical naturalness. In this schema, wilderness is represented as a variable quality, which is entirely appropriate.

The *Wilderness Protection Act* established the WAC.[1] The committee's function was to assess all land in South Australia for protection under the Act. The process for assessment—including reporting, public consultation and decision-making about whether an area of land should be protected under the Act—was mandated by the Act. Protected areas came under the control of the minister for environment and were managed by the director of National Parks and Wildlife, along with other national parks in South Australia.

The WAC had a mandate to assess all lands, regardless of tenure. It worked assiduously to assess existing conservation lands owned by government for their wilderness values, and made recommendations for some of these to change to wilderness areas. These areas were, by and large, intact; most were managed by government within a wilderness management framework.

1   The Wilderness Advisory Committee was abolished in 2015, but its functions were absorbed, unchanged, into the Parks and Wilderness Council under the *National Parks and Wildlife Act 1972* (SA).

108

Early gains on Kangaroo Island in 1993, whereby five wilderness protection areas were proclaimed—Cape Bouguer, Cape Gantheaume, Cape Torrens, Ravine des Casoars and Western River—were not repeated until 2004, when three wilderness protection areas were established on the Eyre Peninsula—at Hambidge, Hincks and Memory Cove. The Billiatt and Danggali Wilderness Protection Areas in the Murray Mallee were proclaimed in 2009, and islands along the west coast were assessed leading to the Investigator Group and Nuyts Archipelago Wilderness Protection Areas being proclaimed in 2011.

There was considerable debate over the land uses that should prevail in wilderness areas. Wilderness is a term with strong connotations. The perception that the land would be 'locked up', ultimately reverting to some 'primitive' long-gone state, prompted concerns about access for recreation and fire management. Compromises had to be reached. Proposals for Avoid Bay and Bascombe Well on the Eyre Peninsula, and Ngarkat in the Murray Mallee, were withdrawn in response to these kinds of concerns. Concerns about the trade-offs that local communities might have to make were largely addressed through explaining how the land would be managed into the future, and offering assurances that people's access to the land would remain fundamentally unchanged.

The 12 wilderness protection areas named above were declared with relatively minimal conflict—the trade-offs were low to non-existent. The areas were all existing no-mining parks with low visitor use, and they were already managed according to wilderness values; it was (without being cavalier) a case of changing the colour on the map from green to dark green. However, this was not the case when the WAC assessed two other conservation areas: Yellabinna and Nullarbor. In both these areas, exploration and mining were permitted within a conservation framework; therefore, it was necessary to consider how to deal with mining as a land use when determining areas that were suitable for wilderness protection.

In 2002, the South Australian Government made a pre-election policy commitment to establish a wilderness area in the Yellabinna wilderness. This vast area of mallee in the far west of the state is covered by over 3 million hectares of parks and reserves. The election commitment followed the previous government's reproclamation of the central portion of the Yumbarra Conservation Park to enable mineral exploration. In undertaking a wilderness assessment of Yellabinna—for the express purpose of identifying an area for protection—the WAC had to consider how to locate a wilderness boundary that would be acceptable to government and allow access for mining in an area that was considered an emerging mineral sands province.

The WAC's initial report established a set of principles for locating and designing a wilderness protection area in the Yellabinna wilderness. The committee called on the relevant government departments (environment and mining) to negotiate a location that met the principles of the report and minimised the impact of future mineral development. This approach proved highly successful. Despite the length of time it took to complete the negotiations, 500,000 hectares of land were eventually identified, which the WAC endorsed. An assessment report recommending the establishment of the Yellabinna Wilderness Protection Area was subsequently prepared, and the area was proclaimed in 2005.

This same process was applied to an assessment of the Nullarbor region. The Nullarbor National Park and Regional Reserve, which provided limited mining access, was assessed and a 900,000-hectare Nullarbor Wilderness Protection Area was established in 2013. This assessment had the added complexity of accommodating visitor access to scenic lookouts along the Nullarbor cliffs.

Both the Yellabinna and Nullarbor processes were highly effective; both relied, to some extent, on strong political commitments to achieve positive outcomes. The process devised by the WAC enabled government to make decisions about land use trade-offs using a set of predetermined principles; however, there was still a degree of conflict regarding these principles. The length of time it took to achieve wilderness protection areas in Yellabinna and Nullarbor caused interest in further wilderness areas to wane. The price of the success of these processes seemed to confirm the view that wilderness and mining were mutually exclusive; indeed, until 2002, creating wilderness areas had nothing to do with mining, as mining was not allowed in conservation lands under assessment.

Wilderness was now at a crossroads. Viewed positively, there was an opportunity for the WAC to take stock of its achievements and consider its future directions. In considering its statutory function to assess all land in South Australia for its wilderness values, the committee considered that it had largely discharged its functions in the settled agricultural areas of the state. However, there were still large areas of the state that had not been assessed. In particular, the arid lands, which make up the majority of the state's landmass, were considered to contain substantial areas of land that would meet the wilderness criteria. Recognising the complexity of tenure and land use in the arid lands, the committee decided to undertake an audit, rather than launch into a new assessment. The committee understood that it was unlikely that a system of wilderness areas could be created in the arid zone, let alone an individual area, for the foreseeable future. It embarked on a process of discussion and debate, reviewed traditional approaches and considered new ways of protecting or managing wilderness. As it rethought the role of wilderness, the review process provoked lively discussions about principles versus pragmatism.

The resulting report, 'Measures for Improving Wilderness Protection in South Australia's Arid Lands', was completed in 2014. The report confirmed that there were areas of high wilderness value and potential national significance in the state's arid area. The diverse tenures in the area—pastoral leases, reserves under the *National Parks and Wildlife Act 1972* and freehold Aboriginal land—prompted the committee to examine wilderness management holistically and to propose a range of measures to protect wilderness values across the vast area. The complex land use and management regimes in the arid lands included:

- management of land for sheep and cattle production
- conservation management by both government and non-government bodies
- land managed for Aboriginal cultural purposes and economic activity
- tourism, transport, defence and mineral and petroleum exploration and extraction.

Against this backdrop, and in consideration of climate predictions for the area, the WAC concluded that while formal protection under the Act remained the preferred means of wilderness protection, in circumstances in which the legislation was not practicable, additional protective mechanisms were required. These additional mechanisms included:

- working with pastoral leaseholders to improve the management of land recognised as having wilderness values, including promoting principles to minimise grazing impacts
- identifying and managing wilderness within parks through existing management plans, without the need to formally excise areas and dedicate them under the *Wilderness Protection Act*

- providing for co-management with Aboriginal people of existing government-owned wilderness protection areas, and enabling the creation of wilderness areas on Aboriginal-owned land[2]
- developing mechanisms to work with Aboriginal people to protect wilderness values on their land.

In addition, the committee recommended that measures should be undertaken to recognise wilderness values in exploration and mining tenements, and to promote better understanding of the significance of arid wilderness and landscape-scale conservation. These recommendations serve as a starting point for further policy work on wilderness; considerably more work is required to translate them into action. The biggest challenge will be reframing wilderness and integrating it into existing and new conservation measures to better balance the perceived land use trade-offs that have become associated with protecting wilderness.

Recommendations

This chapter has raised several issues that warrant further consideration if wilderness is to play a role in contemporary land use planning for conservation outcomes. While the focus has been on South Australia, the issues have relevance nationally.

Wilderness needs to be integrated more fully into protected area methodologies and practices. While it sits comfortably on a spectrum of protected area management categories in the International Union for Conservation of Nature protected area management guidelines, it tends to sit separately in the minds of policymakers, protected area managers and others. As a starting point, it would be useful to examine how park policy managers and field managers perceive wilderness and its role in contemporary Australia.

Wilderness is sometimes perceived as a dated ideal—a 1970s concept about 'saving' areas; this 'saving' discourse often underpins campaigns to create new wilderness areas. It is necessary to reframe this discourse, casting wilderness itself in the role of saviour, for wilderness is critical to

---

2   This has since been enabled through amendments to the *Wilderness Protection Act*; the Nullarbor Wilderness Protection Area has a co-management agreement between the relevant minister and the native title holders.

conserving intact natural areas for refugia and resilience in a changing climate. Affirming the role of wilderness areas as core protected areas to conserve nature efficiently could enable the benefits of such areas to be realised. However, reframing such a value-laden term is no easy matter. Different audiences will react differently; nowhere will this be more challenging and important than with Indigenous Australians, for whom wilderness principles and the injustices of terra nullius are linked.

There is merit in revisiting areas of land that were identified as wilderness areas of potential national significance, as these have driven many of the wilderness assessment priorities. Such a review could determine the current tenure status and wilderness quality of those areas (i.e. have their wilderness values been maintained or diminished?) and revisit the issue of potential national significance by asking new questions about their contribution to climate resilience as intact natural areas.

Finally, following the WAC's lead, it is important to remember that there is opportunity to find ways to leverage existing policy mechanisms, or create new ones, to achieve the protection and sympathetic management of wilderness values on lands where formal government-owned wilderness areas are neither practical nor possible.

8

The Impacts of Land Use Change on Biodiversity in Australia

John Neldner

Key Points

- Habitat loss through land clearing is a leading threatening process of terrestrial biodiversity.
- Increasing intensity of landscape change will increase the loss of wildlife, with a rapid loss once native vegetation falls below 30 per cent.
- As well as the direct impacts on biota of land clearing, the associated fragmentation and habitat modification exacerbates impacts, and there is often an extended extinction debt to be realised.
- Potential economic returns for individuals frequently drives landscape change, whereas the public good from biodiversity services, and the cost of losing these services, are rarely accounted for.
- Legislative and incentive-based approaches are necessary for landscape sustainability.

Introduction

Biodiversity includes all species of organisms, both common and rare. International biodiversity conventions have been signed by many countries. The reasons for preserving biodiversity fall into three general categories: ethics, economics and human welfare (Covacevich, 1995). Regional landscape change can be driven by changes in human population, economic and cultural values, government policy and technology. The magnitude of these changes may be tempered by environmental constraints, but frequently economics is the primary driver (Seabrook, McAlpine & Fensham, 2006).

Intensity of Land Use

Land uses can be regarded as intensive or extensive, depending on the degree to which they modify the environment. More intensive land uses result in a greater share of resources and energy flowing to human uses, leaving less to sustain other species. Hence, different land uses have varying impacts on the biodiversity of an area. The impact on biodiversity increases with the intensity of land use, from no impact to more than 95 per cent loss of the mean species abundance (Taylor, Eber & Toni, 2014). The loss of wildlife species from landscapes tends to occur when clearing exceeds 20 per cent of the landscape, and rapidly accelerates when less than 30 per cent of the native vegetation remains (McAlpine, Fensham & Temple-Smith, 2002; Morgan, 2001). Cautioning against a simple threshold for all landscapes, Maron et al. (2012) have shown that actual thresholds for decline may be affected by landscape productivity and the natural cover of the vegetation. Land cover has been used as a surrogate for habitat for terrestrial biodiversity, and disturbed land cover represents a deleterious change in the habitat suitability (Graetz, Wilson & Campbell, 1995). Morgan (2001) mapped landscape health in Australia by integrating the state and trend in native vegetation extent, connectivity and condition, dryland salinity, hydrology, weeds, feral animals, threatened ecosystems and species. Broadly, Australia was divided into two zones: the intensive use zone, represented by large areas of fragmented land and significant areas of cropping and intensive domestic grazing, and the extensive use zone, where very little clearing had occurred and extensive grazing was the predominant land use. Thackway and Lesslie (2006) classified vegetation into six broad condition classes ranging from natural systems through to completely transformed environments. While these frameworks are useful tools for communicating the current condition and trend of vegetation and landscapes, Eyre, Fisher, Hunt and Kutt (2011) argued that they are a poor indicator of biodiversity persistence and trend, and that what is required is a comprehensive monitoring framework to measure biodiversity and driver indicators, both directly and indirectly.

Direct Effects on Biodiversity

Land clearance is one of 16 threatening processes recognised for Australian terrestrial biodiversity under the *Environment Protection and Biodiversity Conservation Act 1999* (EPBC), and is regarded as exerting one of the most intense impacts on biodiversity. Using accurate tree clearing mapping (Accad, Neldner, Wilson & Niehus, 2001) and estimates of average densities of tree and vertebrate animal species in broad vegetation types, Cogger, Ford, Johnson, Holman and Butler (2003) conservatively estimated that, in the Brigalow Belt of Queensland, more than 112 million trees were cleared each year between 1997 and 1999. They further estimated that this resulted in the deaths of 1 million mammals, 5 million birds and 52 million reptiles. Not all these animals were killed directly by the clearing process—some mobile species would have escaped to remnant vegetation; however, they would have soon perished from starvation or predation in these fully occupied and frequently fragmented and degraded habitats. McAlpine et al. (2002) estimated that a reduction in remnant vegetation to 30 per cent would result in the loss of 25–35 per cent of the vertebrate fauna; however, the full impact may take more than 100 years. While habitat loss is the primary impact of clearing on biota, the fragmentation and modification of the remaining habitat has a strong secondary deleterious effect (Haddad, Brudvig & Clobert, 2015). Patch size and landscape connectivity have been shown to have a strong relationship with retained biodiversity (Bowen, McAlpine, House & Smith, 2007).

Indirect Effects

The impact of vegetation clearance on biodiversity may take decades to become apparent (Cogger et al., 2003; McAlpine et al., 2002). Extinction debt works through local extinctions gradually becoming regional: eventually, the entire species is made extinct. In addition, immigration lag, in which small or isolated patches are slower to accumulate species, has been shown to result in 5 per cent fewer species after one year, and 15 per cent fewer species after 10 years (Haddad et al., 2015). The third process of degradation caused by fragmentation is ecosystem function debt. Haddad et al. (2015) recorded decreases in nutrient cycling and plant and consumer biomass of up to 80 per cent after 10 years in small fragments. It is difficult to measure these changes and disaggregate their causes, as frequently there are many interactions through processes such as habitat fragmentation and degradation, contact with non-native species and flow-on effects from outside the locality (e.g. extreme weather associated with climate change) that lead to decreased resilience.

Twelve threatening processes listed under the EPBC for terrestrial Australia are based on exotic plants (e.g. gamba grass, escaped garden plants and novel biota) or animals (e.g. rabbits, goats, foxes, cats, pigs, fire ants and cane toads), three are based on the effects of exotic diseases and one is based on a native species (i.e. noisy miner, *Manorina melanocephala*). Within the suite of species that occupies a given location, there may be some winners (increasers) who benefit from the new environmental conditions provided by clearing (e.g. large kangaroos), while other species will be losers (decreasers) or neutral (e.g. small woodland birds). Feral animals, such as cats and foxes, directly prey on Australian biota, while rabbits, goats and pigs degrade the landscape, consume native plants and directly compete with native herbivores for food. Exotic plants can outcompete native plants (e.g. linear declines in plant species richness as buffel grass cover increases) (Fensham, Wang & Kilgour, 2015), or create conditions that transform landscapes and make them unfavourable for native species (e.g. gamba grass produces high biomass loads that encourages tree-killing hot fires in the tropical savannas). The clearing, or thinning, of vegetation may lead to a native species becoming a competitive excluder of other biodiversity; for example, yellow-throated miners dominating and excluding smaller birds (<53 grams) throughout 500,000 square kilometres of northern Australian rangelands (MacNally et al., 2014). The non-native fungal disease *Phytophthora cinnamomi* has led to the degradation of biodiversity in the Western Australian heathlands, particularly affecting the family Proteaceae. Myrtle rust has a similar potential in the east-coast rainforests.

While land clearing causes a substantial loss of biodiversity, significant negative impacts can occur where only subtle land use change has been recorded. In Australia, the very high rate of land mammal extinction—10 per cent in the last 200 years—is most likely due to predation by feral cats and foxes and changed fire regimes (Woinarski, Burbidge & Harrison, 2015). Large areas of intact vegetation in the extensive land use zone of northern Australia, and even in protected areas such as Kakadu National Park, are exhibiting dramatic mammal losses. To put this in perspective, in 1996, small mammals were captured in 89 per cent of quadrats in Kakadu National Park, compared to only

35 per cent in the 2000 sampling. There is considerable interest in the reintroduction of the dingo as an apex predator to assist in the restoration of rangelands through controlling feral predators (Newsome et al., 2015).

Assessing and Measuring Impacts

We can understand the changes to biodiversity through inventories and comparison of analogous ecosystems in different condition states or, more thoroughly, through long-term monitoring of ecosystems undergoing land use change. Robust assessments of the condition of the Australian environment will only be possible if a representative sample of ecosystems is monitored using stratified and replicated plots over the long term (Eyre et al., 2011; Lindenmayer, Burns & Tennant, 2015). The monitoring of key components of ecosystems needs to commence as soon as possible and continue for at least several decades to provide information to guide ecologically sustainable development that retains biodiversity. Listing species—and, increasingly, ecological communities—as threatened has been seen as a gauge of the impact of land use change on biodiversity; however, this is an uncertain gauge, as we have very limited knowledge for many species. The International Union for Conservation of Nature (2012) criteria for assessment of threatened species examines population decrease 'over the last 10 years or three generations, whichever is the longer time period' to determine decline and threat. In many cases, the major reduction in habitat and species population occurred many years earlier; consequently, the assessment is based on decline within the depleted population, rather than on what the species population would have been prior to clearing. For example, the brigalow ecological community and many brigalow-dominated regional ecosystems have experienced huge losses of remnant habitat and have been listed as endangered under the EPBC and Queensland's *Vegetation Management Act 1999* (VMA); however, the number of listed threatened or near-threatened species living under Queensland's *Nature Conservation Act 1992* (NCA) in these areas is low—only eight species of reptiles, two mammals, six birds, one butterfly and three vascular plants (see Figure 8.1). Hence, the threatened species list is a blunt indicator of the status of individual species as compared to pre-European populations, and of landscape change impacts on biodiversity.

Ameliorating the Impacts? Can Biodiversity and Development Survive Together?

While conservation within an expanded reserve system remains an important priority, many native fauna species occupy modified landscapes outside the formal reserve system. Further, the reality for agricultural landscapes suffering high levels of habitat loss and fragmentation is that the protection of remnant (i.e. not previously cleared) habitat alone will not be enough to achieve biodiversity conservation goals, and some form of landscape restoration will be necessary (Bowen et al., 2007).

Restoration of landscapes (e.g. rehabilitation plantings) can improve biodiversity values, but there may be a considerable lag before suitable habitat requirements are met; for example, hollows in large trees take more than 100 years to develop. Native woody plants can colonise extensive areas when there are social drivers, such as the change from traditional agriculture to rural amenity use, resulting in 1,800 hectares per decade becoming shrublands on low fertility soils in central Victoria since the 1960s (Geddes, Lunt, Smallbone & Morgan, 2011). This old field succession, or natural regrowth, can be far more extensive than intentional plantings (Fensham & Guymer, 2009).

Landscape change is driven by economic, demographic and cultural factors operating at a range of spatial and temporal scales, with the drive to maximise the return on an investment often being most important (Graetz et al., 1995; Seabrook et al., 2006). Since it is very difficult to control the economic market, governments attempt to exert influence through planning and regulation (and compliance) and, occasionally, through the use of incentives. The concept of ecosystem services and the market providing a positive incentive to retain values is potentially starting to occur through the carbon market; however, it is not operational for most biodiversity values and services (e.g. pollination from flying foxes). A price on carbon is driving land use changes globally, but the benefit to biodiversity depends on the planting methodology used (Carwardine et al., 2015; Fensham & Guymer 2009). Other major political drivers, such the Carbon Farming Initiative and offsets policies, attempt to drive landscape change and compensate for biodiversity loss in development areas. Biodiverse plantings that attempt to restore pre-clearing vegetation by including a full suite of plant species and lifeforms will generally produce more biodiversity benefits, and be more resilient to climate variability and fire impacts (Dwyer, Fensham, Butler & Buckley, 2009; Fensham & Guymer, 2009). The biodiversity contained in rehabilitation depends on many factors, including the ability of species to move to the new habitat, soil and seasonal conditions to allow germination and follow-up management to allow successful establishment and reproduction for self-sustaining populations. Higher biodiversity benefits would be expected from restoration plantings in endangered ecological communities, habitats for threatened species and areas well connected to other natural vegetation (Carwardine et al., 2015; Munro et al., 2009).

A Case Study of Brigalow Communities

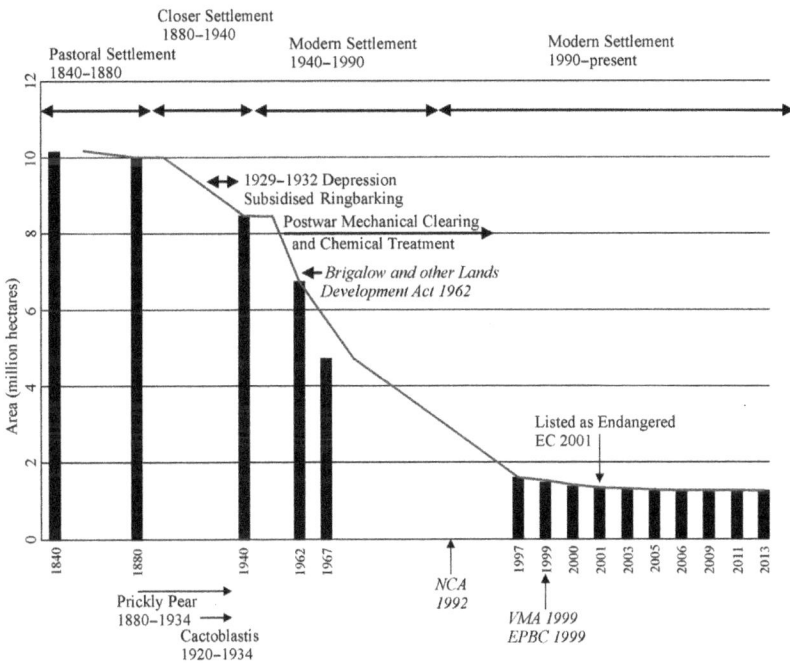

Figure 8.1: Diagram illustrating the decline in extent of the brigalow broad vegetation group.

Source: Neldner et al. (2017). Annotated with social, economic and legislative drivers and extent figures derived from Seabrook et al. (2006). Pre-clearing and remnant brigalow extent 1997–2013 from Accad & Neldner (2015).

Figure 8.1 summarises the range of social, economic, legislative and environmental factors described in Seabrook et al. (2006) that have affected the brigalow ecological community. Since European settlement, the brigalow ecological community has declined from 10.2 million hectares (5.9 per cent of Queensland) to be an EPBC-listed Endangered Ecological Community of only 1.3 million hectares in 2011 (Neldner et al., 2017). A recovery plan developed in 2001 included protection of the remnant patches in the conservation estate, protection of regrowth older than 15 years under the EPBC (and sometimes the VMA under various conditions) and the potential enhancement of some regrowth through offsets. The remaining brigalow is likely to be further fragmented via infrastructure for coal seam gas extraction and processing; it is also threatened by the encroachment of non-native species, such as buffel grass, that can lead to fire regimes that degrade the community; for example, at Mazeppa National Park (Butler & Fairfax, 2003). However, eucalypt and brigalow woodlands are the most likely to benefit from reforestation planting because of the relative cost effectiveness of carbon sequestration in those systems (Carwardine et al., 2015).

Recommendations

1. The impact of land use change on biodiversity is complicated and includes direct and indirect impacts. Continued systematic monitoring of both flora and fauna at both the landscape scale (including changes in extent and condition) and site scale (multiple attribute) is required to understand and manage these impacts.

2. A number of approaches (both legislative- and incentive-based) are required to ensure that what biodiversity remains in fragmented (e.g. brigalow) and even largely unaltered landscapes (e.g. savannas) is retained and sustainably managed.

References

Accad, A. & Neldner, V. J. (2015). *Remnant vegetation in Queensland: Analysis of pre-clearing, remnant 1997-1999-2000-2001-2003-2005-2007-2009-2011-2013 regional ecosystem information.* Brisbane, QLD: Queensland Herbarium, Department of Science, Information Technology and Innovation.

Accad, A., Neldner, V. J., Wilson, B. A. & Niehus, R. E. (2001). *Remnant vegetation in Queensland: Analysis of pre-clearing, remnant 1997–1999 regional ecosystem information*. Brisbane, QLD: Queensland Herbarium, Environmental Protection Agency.

Bowen, M. E., McAlpine, C. A., House, A. P. N. & Smith, G. C. (2007). Regrowth forests on abandoned agricultural land: A review of their habitat values for recovering forest fauna. *Biological Conservation 140*, 273–96. www.sciencedirect.com/science/article/pii/S0006320707003308

Butler, D. W. & Fairfax, R. J. (2003). Buffel grass and fire in a gidgee and brigalow woodland: A case study from central Queensland. *Ecological Management and Restoratio*n *4*, 120–25. doi.org/10.1046/j.1442-8903.2003.00146.x/abstract

Carwardine, J., Hawkins, C., Polglase, P., Possingham, P., Reeson, A., Renwick, A. R., … Martin, T. (2015). Spatial priorities for restoring biodiverse carbon forests. *Bioscience 65*(4), 372–82. doi.org/10.1093/biosci/biv008

Cogger, H. G., Ford, H. A., Johnson, C. N., Holman, J. & Butler, D. (2003). *Impacts of land clearing on Australian wildlife in Queensland*. Brisbane, QLD: World Wildlife Fund Australia.

Covacevich, J. A. (1995). Realities in the biodiversity Holy Grail: Prospects for reptiles of Queensland's brigalow biogeographic region. *Proceedings of the Royal Society of Queensland 106*, 1–9.

Dwyer, J. M., Fensham, R. J., Butler D. W. & Buckley, Y. M. (2009). Carbon for conservation: Assessing the potential for win-win investment in an extensive Australian regrowth ecosystem. *Agriculture Ecosystems and Environment 134*, 1–7. doi.org/xml/10.2307/25741351

Eyre, T. J., Fisher, A., Hunt, L. P. & Kutt, A. S. (2011). Measure it to better manage it: A biodiversity monitoring framework for the Australian rangelands. *The Rangeland Journal 33*, 239–53. doi.org/10.1071/RJ10071

Fensham, R. J. & Guymer, G. P. (2009). Carbon accumulation through ecosystems recovery. *Environmental Science and Policy 12*, 367–72. doi.org/10.1016/j.envsci.2008.12.002

Fensham, R. J., Wang, J. & Kilgour, C. (2015). The relative impacts of grazing, fire and invasion by buffel grass (*Cenchrus ciliaris*) on the floristic composition of a rangeland savanna ecosystem. *The Rangeland Journal 37*, 227–37. doi.org/10.1071/RJ14097

Geddes, L. S., Lunt, I. D., Smallbone, L. T. & Morgan, J. W. (2011). Old field colonization by native trees and shrubs following land use change: Could this be Victoria's largest example of landscape recovery? *Ecological Management and Restoration 12*, 31–6. doi.org/10.1111/j.1442-8903.2011.00570.x/abstract

Graetz, R. D., Wilson, M. A. & Campbell, S. K. (1995). *Landcover disturbance over the Australian continent—a contemporary assessment* (Biodiversity Series, no. 7). Canberra, ACT: Department of the Environment, Sport and Territories.

Haddad, N. M., Brudvig, L. A. & Clobert, J. (2015). Habitat fragmentation and its lasting impact on Earth's ecosystems. *Science Advances 1*(2). doi.org/10.1126/sciadv.1500052

International Union for Conservation of Nature. (2012). *IUCN red list categories and criteria: Version 3.1* (2nd ed.). Cambridge, UK: IUCN.

Lindenmayer, D. B., Burns, E. L. & Tennant, P. (2015). Contemplating the future: Acting now on long-term monitoring to answer 2050's questions. *Austral Ecology 40*, 213–24. doi.org/10.1111/aec.12207

MacNally, R., Kutt, A. S., Eyre, T. J., Perry, J. J., Vanderduys. E. P., Mathieson, M. ... Thomson, J. R. (2014). The hegemony of the 'despots': The control of avifaunas over vast continental areas. *Diversity and Distributions 20*, 1071–83. doi.org/10.1111/ddi.12211

Maron, M., Bowen, M., Fuller, R. A., Smith, G. C., Eyre, T. J., Mathieson, M., ... McAlpine, C. A. (2012). Spurious thresholds in the relationship between species richness and vegetation cover. *Global Ecology and Biogeography 21*, 682–92. doi.org/10.1111/j.1466-8238.2011.00706.x

McAlpine, C. A., Fensham, R. J. & Temple-Smith, D. E. (2002). Biodiversity conservation and vegetation clearing in Queensland: Principles and thresholds. *The Rangeland Journal 24*, 36–55. doi.org/10.1071/RJ02002

Morgan, G. (2001). *Landscape health in Australia. A rapid assessment of the relative condition of Australia's bioregions and subregions.* Canberra, ACT: Environment Australia and National Land and Water Resources Audit.

Munro, N. T., Fischer, J., Wood, J. & Lindenmayer, D. B. (2009). Revegetation in agricultural areas: The development of structural complexity and floristic diversity. *Ecological Applications 19*(5), 1197.

Neldner, V.J., Niehus, R.E., Wilson, B.A., McDonald, W.J.F., Ford, A.J. & Accad, A. (2017). *The Vegetation of Queensland. Descriptions of Broad Vegetation Groups. Version 3.0.* Queensland Herbarium, Department of Science, Information Technology and Innovation, Brisbane. Retrieved from publications.qld.gov.au/dataset/redd/resource/78209e 74-c7f2-4589-90c1-c33188359086

Newsome, T. M., Ballard, G., Crowther, M. S., Dellinger, J. A., Fleming, P. J. S., Glen, A. S., … Dickman, C. R. (2015). Resolving the value of the dingo in ecological restoration. *Restoration Ecology 23*, 201–08. doi.org/10.1111/rec.12186

Seabrook, L., McAlpine, C. & Fensham, R. (2006). Cattle, crops and clearing: Regional drivers of landscape change in the Brigalow Belt, Queensland, Australia, 1840–2004. *Landscape and Urban Planning 78*, 373–85. doi.org/10.1016/j.landurbplan.2005.11.007

Taylor, M. F. J., Eber, S. C. & Toni, P. (2014). *Changing land use to save Australian wildlife.* Sydney, NSW: World Wildlife Fund Australia.

Thackway, R. & Lesslie, R. (2006). Reporting vegetation condition using the vegetation assets, states and transitions (VAST) framework. *Ecological Management and Restoration 7*, S53–S62. doi.org/10.1111/ j.1442-8903.2006.00292.x

Woinarski, J. C. Z., Burbidge, A. A. & Harrison, P. L. (2015). Ongoing unravelling of a continental fauna: Decline and extinction of Australian mammals since European settlement. *PNAS 112*(15), 4531–40. doi.org/10.1073/pnas.1417301112

Part 3 – Working to Achieve National Coordination

9

National Coordination of Data and Information to Inform Land Use Policies and Programs: The Recent Past, the Present and Ideas for the Future

Richard Thackway

Key Points

- The Australian Government has responsibility for the development of national coordination arrangements for land use–related natural resource management (NRM) data and information to inform national land use policies and programs.

- If Australia wants a nationally coordinated approach to land use and land management, the Australian Government must provide consistent leadership and support—it is a matter of political and national will and mandate.

- So-called fundamental or core land use–related NRM national datasets are not immutable; as government NRM and land use policies change, so does the need for new or different fundamental datasets.

- Effort should be invested in defining, collecting and improving data repositories that are based on essential environmental measures rather than derived environmental datasets. The later can be generated as required from the former.

- Enduring and widely used national NRM datasets, i.e. those that have been adopted and maintained for longer than 10 years, share similar characteristics (Box 9.1).

- National decision-makers require access to a wide variety of NRM data and information to improve land use policies and planning—for example, environmental flows. The greater the area being targeted, assessed or monitored, the greater is the need for national coordination and governance arrangements—for example, the Murray-Darling Basin Authority.

Introduction

The Australian Government has responsibility for the development of national coordination arrangements for land use related natural resource management (NRM) data and information to inform land use policy. This chapter is written from the perspective of a research scientist who was embedded in numerous science policy units in the Australian Government from 1984 to 2011. Over that period, I was responsible for developing some key NRM data and information products that were subsequently used to inform science policy and influence NRM programs, and were widely recognised for their contribution to informing public–private decision-making. These products include bioregions (Thackway & Cresswell, 1995), Indigenous protected areas (Thackway, Szabo & Smyth, 1997), weeds of national significance (McNaught, Thackway, Brown & Parson, 2008), revegetation (Atyeo & Thackway, 2009), native vegetation condition (Thackway & Lesslie, 2008) and dynamic land cover (Thackway, Lymburner & Guerschman, 2013). This role had its challenges, such as:

- producing timely, scientifically credible and policy-relevant advice and information, while keeping abreast of rapidly changing technological and scientific developments

- developing agreed and enduring natural resource data and information products that (to the extent possible) were neither partisan to those working in biodiversity conservation and protection, nor to those involved in sustainable land use and management

- engaging with, and transcending, various public service cultures, including those who regarded the states and territories as a hindrance to developing consistent national data and information; those who believed that cutting budgets would not compromise development of the same or better-quality data and information products; and those who opposed scientists publishing their science policy–relevant work in the scientific literature, thereby restricting the development of their professional profile and standing.

Against these challenges, the following observations reflect my deep understanding of the characteristics of coordinated national land use–related policy and planning, NRM data and information; how these products are developed and maintained through partnerships; and how they are used in land use policy and planning and public programs at local, regional, state and national levels.

There has been an evolution in the operation of Australian government agencies and their relationships with data suppliers, particularly the states and territories, since the early 1980s. Five broad phases of national coordination can be recognised:

- Phase 1—before 1980: there was limited cross-border coordination between the states and territories. States and territories operated independently and were responsible for land use and management and developing natural resource data and information coordination and assessment programs.
- Phase 2—1980–99: the states and territories had significant natural resource data and information coordination and assessment programs and the Australian Government's national coordination was in its formative stage. National coordination had to be extensively promoted (e.g. Working Group for Land Resource Assessment).
- Phase 3—2000–07: the Australian Government sponsored and supported partnerships and bilateral agreements with the states and territories through national natural resource data and information coordination committees (e.g. those reporting to the Advisory Committee of the National Land and Water Resources Audit [NLWRA], the National Committee on Soil and Terrain, the Executive Steering Committee for Australian Land Use Mapping and the Executive Steering Committee for Australian Vegetation Information).

These committees produced nationally agreed protocols and major national assessments that contributed to informing discussions on land use related policies.

- Phase 4—2008–13: the Australian Government dramatically reduced its role in sponsoring most national natural resource coordination organisations and committees; this included major reductions in funding to support collaboration with state land management agencies. Instead, it invested in the development of national natural resource data infrastructures (i.e. Australia-wide datasets) by directly funding research agencies and universities (e.g. the National Collaborative Research Infrastructure Strategy, Atlas of Living Australia [NCRIS ALA] and Terrestrial Ecosystem Research Network [TERN]). The states reviewed and revised their investments in national natural resource data and coordination with the Australian Government.

- Phase 5—2014 to present: the Australian Government has experienced significant declines in revenue. As a consequence, land use–related federal agencies have redefined their roles and responsibilities, including curtailing or terminating national coordination activities.

Numerous factors have contributed to this evolution, including:

- states and territories have not invested in the collection of new natural resource management (NRM) data and information since Phase 2
- significant advances in the speed of computing and the decreasing costs of computers and computer storage
- development of data infrastructure facilities that support major archives of spatial and temporal data and information (e.g. TERN)
- development of citizen science and online facilities to support standardised collection of field data and rapid connections between individuals and data warehouses (e.g. NCRIS ALA)
- development of more sophisticated modelling and scenario tools that are designed to ingest and analyse large multi-temporal image data archives (e.g. TERN facilities)
- growth of handheld personal communication tools and social media that enable individuals to collect, store, access, upload and download data and information from national data repositories
- growth in the legislative and regulatory powers of Australian government agencies and associated budgets that support data acquisition, data warehousing, analytics and internet access and reporting

- growth in metadata systems, including the Australian Spatial Data Directory, which provide a national metadata hub for searching other national, state and territory directories to facilitate the discovery of published geospatial datasets throughout Australia.

Over recent years, the Australian Government's capacity to engage in coordination of land use–related policy and planning has been significantly reduced. One likely outcome of reduced budgets and functions is that the Australian Government may revise its former status as national coordinator for land use policy and planning (as described in Phases 1–2), which, given that the Australian Constitution vests responsibility for land use and management with the states and territories, would be justified.

If Australia wants a nationally coordinated approach, the Australian Government must provide leadership and support: it is a matter of political and national will and mandate. The Constitution aside, the Australian Government has responsibility for cross-jurisdictional issues of national and international significance. Moreover, it has signed various international treaties and conventions that carry responsibilities for monitoring and reporting. However, it must choose to exercise these responsibilities. Bilateral agreements have proven useful in the past; however, simply telling the states and territories that they are individually responsible for land management and NRM does not, itself, provide a nationally coordinated response.

Data Needs Identification

The Australian Government supports the acquisition of a wide array of biophysical, socio-economic and NRM-related data, and coordinates the development of information products for numerous land use and planning purposes. Most of these products require access to up-to-date, spatially accurate and policy-relevant data so that the information can be appropriately used to support policy development and improve decision-making.

Figure 9.1 shows a generic conceptual model that provides a convenient adaptive management framework for guiding the fundamental data and information items that are required—when, where and at what level of spatial and temporal detail (Thackway et al., 2013). This model, which has five key decision stages, has been used extensively in a science policy context (e.g. Thackway et al., 2013). For the purposes of illustration, the focus is on how the model can be used to determine the ecosystem services that are required by Australian and state government agencies, regional bodies and land managers.

This 'stepped cycle' model provides a framework that can be used for clarifying and addressing issues related to the what, why, how, when and where of future national resource management programs—specifically, how they can deliver better land use policy and program outcomes. The framework demonstrates how issues of scale of data can best be understood, and how this information can be used at each step of a strategic decision-making approach at different levels (national, regional and local). The key decision points provide useful checkpoints for reviewing and evaluating the appropriateness and relevance of data and information before proceeding to the next point. The model highlights both the gaps and need to collect new data before progressing to the next point. It is based on the premise that decisions should be supported by a clear appreciation of the data, information needs and priorities, sound understanding of the availability of suitable resources and options for their use, and capacity to measure, monitor and report changes in on-ground attitudes and support for land use and land management practices.

This model can be repeated in progress towards long-term objectives or, as necessary, in response to changing environmental conditions or policy and program priorities, and can be applied to different stakeholder groups operating at different spatial and temporal scales, such as public policy and program managers (e.g. federal and state governments), regional bodies (e.g. catchment management authorities) and land managers. There are interactions and crossovers between the different stakeholder groups. Collectively, these decision-makers may cooperate to deliver improved land use outcomes through adaptive management. Figure 9.1 illustrates these interactions with varying spatial and temporal scales, and is accompanied by a corresponding set of five broad decision points.

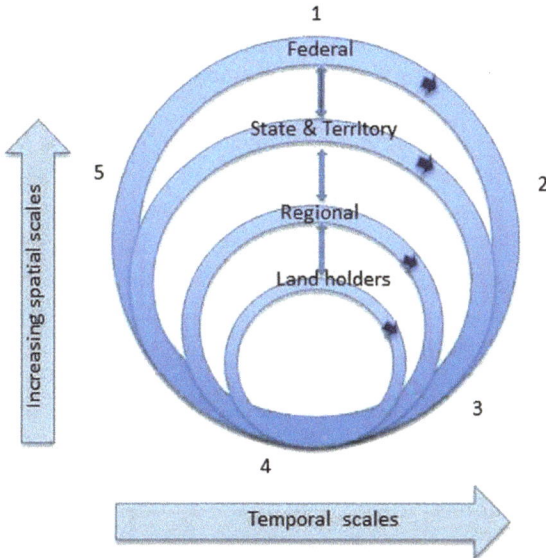

Steps or decision points for managing land use–related NRM outcomes

Step 1 Asset definition:

- Determine the appropriate landscape scale, characterise the mosaic of land use types and their ecosystem function (when and where).

Step 2 Identify land use characteristics:

- Determine the extent that the required ecosystem services are supplied by the current land use types and their ecosystem function and assess how the socio-ecological setting supports or limits their capacity.

Step 3 Identify needs for change:

- Determine if (and where in the landscape) changes in land management practices will maintain or enhance the condition of assets and hence improve the mix of ecosystem services.

Step 4 Identify and select options and implement priority actions:

- Set priorities for actions, consider trade-offs involved and identify areas for intervention whereby actions are to be undertaken through existing, revised or new policy and programs or changes in land management practices; invest in interventions that match selection criteria and monitor land use and land cover responses and links to ecosystem services and the effects of investments; and integrate relevant monitoring data with existing database systems.

Step 5 Evaluate the responses of the land cover to changes in land management practices:

- Analyse the spatial and temporal patterns and analyse how well the land use outcome met the desired goals and targets; repeat Steps 1–5 as required.

Figure 9.1: Steps in assessing data and information required for science-policy applications.

Source: Thackway, Lymburner & Guerschman (2013, modified from Figure 2).

Figure 9.1 also shows the conceptual process for developing datasets that involve consideration of multiple spatial and temporal scales and various stakeholders; however, in practice, the situation is rarely this straightforward. Experience in developing regional-scale nationally consistent datasets that have been used by policymakers and planners for more than 10 years involves a complex interplay of applied science and research, federal–state and public–private relationship building, publication and marketing, enhancements and continuous improvement. Box 9.1 sets out the characteristics of enduring datasets that are based on sound national cooperation and collaboration, usually in partnership with the states and territories.

Box 9.1: Characteristics of enduring datasets[1] that are based on sound national cooperation and collaboration.

1. Address a well-defined problem, issue or key questions
2. present an agreed conceptual model, framework or information hierarchy
3. have an effective champion, sponsor or leader
4. are underpinned by unambiguous governance arrangements
5. have a strong relevance to one or more policies and programs
6. are developed and maintained by recurrent adequate resourcing (people and ~ $)
7. are underpinned by sound technical, scientific and information technology (IT) support
8. are underpinned by sound data management enabling interoperability and a capacity to integrate
9. have sound data models enabling flexible information products to be generated
10. have been published and peer reviewed
11. are supported by custodians committed to continuous improvement (spatial and temporal)
12. are supported by a continuum of levels of detail, processing and standards
13. are discoverable, reusable and accessible
14. are relevant to research and education
15. are relevant to planners and on-ground managers
16. are relevant to key clients or partners
17. represent the ideal 'the whole is more than the sum of its parts' (i.e. products add value to the inputs from jurisdictions).

---

1   Regional-scale national datasets that are used by policy and planning for more than 10 years.

Influence Monitoring through Developing and Promoting Core National Attributes

Beyond its importance for design and implementation of policies and programs, national coordination of key land use–related NRM data and information has been vital to the monitoring, evaluation, reporting and improvement of policies and programs. While millions of dollars have been invested in public land use–related NRM policies and programs, a review of the Natural Heritage Program in 2008 was unable to assess whether existing management interventions had solved the environmental problems related to adverse effects of land use and land management, or whether the investment had been cost-effective (Australian National Audit Office, 2008). The critical issue was gaining access to the up-to-date, detailed spatial and temporal information necessary to ascertain how successful the interventions were, given their apparent need.

Evidence-based land use–related NRM policy and program settings that are founded on credible spatial and temporal data and information can make a stronger case for early interventions, renewed funding and sound evaluations of performance. The development of nationally consistent regional-scale mapped datasets is commonly based on the relevant Australian government agency working with appropriate state and territory land management agencies to develop protocols and supporting datasets. The process is characterised by the Australian Government initiating, sponsoring and sometimes funding cooperative and collaborative projects under an appropriate executive science policy governance arrangement. The period between 2000 and 2005 saw the rapid development of online data collection and mapping tools that, combined with high-speed data transfer developments, were quickly accepted and promoted by land use policy and program managers as platforms for improving the efficiency and effectiveness of program delivery and performance monitoring.

The control and management of weeds of national significance (WONS) and the need to use revegetation to address a wide range of NRM-related issues arising from over-cleared landscapes are two cases of national coordination of core attribute data. Core attributes represent the minimum number of features that are required in surveying, mapping, monitoring and reporting across different scales. The obvious advantage of using standardised national core attributes is that new data collected using these protocols are more likely to be spatially consistent across scales and over time, and existing data can be transposed where they fit the attribute minimum standards.

Core attributes have been developed for WONS by McNaught et al. (2008) and for revegetation by Atyeo and Thackway (2009) through extensive consultation with national, state and regional program managers and on-ground local-scale project practitioners. High-level councils contributed to the recognition of core attributes for use in state and national public programs; WONS involved the Australian Weeds Council and revegetation involved the NLWRA.

Core attributes have been used to improve design, delivery and performance of NRM programs at the strategic and tactical levels. At the strategic level (i.e. regional, state and national scales), attributes describing the type, extent and distribution of weeds and revegetation are required for range of purposes. Monitoring at this scale is sometimes described as 'surveillance monitoring'. The WONS surveillance maps were used to design and implement the Australian Government's Defeating the Weed Menace program by targeting areas for control of new weed incursions. Surveillance maps of over-cleared landscapes were used to target revegetation programs, including Landcare, Bushcare, Rivercare, the Natural Heritage Trust, Corridors of Green, Greening Australia, Envirofund and Caring for our Country. At the tactical level, core attributes have also been influential in designing on-ground projects and documenting the outcomes of management activities. Monitoring at this scale is described as 'investigative monitoring'.

Informing Land Use Debates Using High-Quality Fundamental Datasets

Scientifically based information, or fundamental data, that everyone can agree on and trust should underpin land use and land management debates. Information about land use is especially important for better management of natural resources. The role of a data provider delivering these fundamental datasets is a critical component of the debate. The provider must be trusted and the information must have a well-documented pedigree to ensure its integrity (i.e. metadata). The general community, industry and developers share an interest in and require access to information sets. The Australian Government has been a provider of such land use datasets information under the auspices of the Australian Collaborative Land Use and Management Program (ACLUMP, 2010). Membership of ACLUMP includes Australian, state and territory government partners. The Australian Land Use and Management Classification has been adopted by all partner agencies and is supported by nationally agreed technical standards.

Commencing in 2008, ACLUMP made Australia-wide land use mapping available at national and regional (catchment) scales. National-scale maps (1:2,500,000) and data are available online for a series of years from 1992–93 to 2005–06. Catchment and regional maps are available at a variety of scales—1:25,000 to 1:100,000—except in sparsely settled areas, where the recommended scale is 1:250,000. The currency of catchment-scale mapping ranges from 1997 to 2015.

Land use data provide context that is important, and often essential, to meeting the information needs for monitoring, evaluation and reporting against indicators used by natural resource management programs. Land use–related NRM datasets are variously used for policy initiatives, including establishing opportunities and barriers for resource development, populating particular indicator frameworks (State of the Environment [SoE], 2011), contributing to long-term records (e.g. bushfire and climate databases) or developing scenarios of the projected extent and condition of natural resources and environmental variables under different land use futures. Similarly, baseline spatial and temporal information on land management practices is essential for monitoring and reporting on progress towards long-term resource condition outcomes.

On a global scale, land use–related issues have been shown to have significant effects on protective functions and ecosystem services of global forests (Miura et al., 2015). At this scale, land use and how the land is managed have obvious effects on the condition of native vegetation and consequences for biodiversity (Thackway, 2016). On a national scale, evidence has shown that, by combining land use data and information with other NRM datasets (e.g. land salinity) and native vegetation (NLWRA, 2007, 2008a, 2008b), key questions can more readily be answered, such as:

- What is the nature and extent of the issue and how does it relate to land use?
- Is the existing or proposed land management intervention appropriate for the size of the issue?

- What types of land management intervention work best, are most cost-effective and have the best transferability across regions?
- What was the impact of the land use–related policy or program investment—in the intermediate and long term?

Monitoring and evaluation of core indicators support evidence-based decision-making at national, state, territory, regional and land manager levels (see Figure 9.1). Three national examples in which land use–related NRM data and information are critical for regular national monitoring and reporting are the national SoE (2011), state of the forest reports (Montreal Process Implementation Group for Australia and National Forest Inventory Steering Committee, 2013) and regional environmental accounting (Sbrocchi et al., 2015). Figure 9.1 acknowledges that decision-makers may have a wide variety of data and information needs in terms of content, context or spatial and temporal scales at each level. Equally, there is complexity across these four levels due to multiple needs, values, preferences and time frames.

As noted above, we have seen the waxing and waning of national coordination since the late 1970s. The demise of key coordination agencies, such as Land and Water Australia and NLWRA and the diminution of support for most national coordinating committees have created a vacuum in national coordination of data and information to inform land use policies and programs. Given the critical nature of land use and land management data to land use decision-making, clearly much more needs to be done. National coordination is needed in the following key areas:

- identifying fundamental datasets to support key agency policies and programs, such as Australian Bureau of Agricultural and Resource Economics and Sciences (Mutendeudzi & Stafford-Bell, 2011)
- identifying key collection agencies and funding them to collect fundamental data, as proposed in the National Plan for Environmental Information (Australian Government Environmental Information Advisory Group, 2012)
- clarifying the future role of citizen science in collecting land use and management data
- supporting the development of, and investing in, essential environmental measures that underpin national datasets, which are critical for land use decision-making, including monitoring and reporting, program design and evaluation

- clarifying whether we should legislate data collection and transfer (e.g. *the National Water Act 2007*)
- clarifying what agency at the federal level has carriage for undertaking sustainability assessments (e.g. the Productivity Commission).

Acknowledgements

Chris Auricht, Shane Cridland, Peter Wilson, Blair Wood and Graham Yapp provided comments on an earlier draft.

References

ACLUMP (Australian Collaborative Land Use and Management Program). (2010). *Land use and land management information for Australia: Workplan of the Australian Collaborative Land Use and Management Program.* Canberra, ACT: Australian Bureau of Agricultural and Resource Economics and Sciences.

Atyeo, C. & Thackway, R. (2009). Mapping and monitoring revegetation activities in Australia—towards national core attributes. *Australasian Journal of Environmental Management 16*(3), 140–48. Retrieved from search.informit.com.au/document Summary;dn=071365261532926;res=IELBUS

Australian Government Environmental Information Advisory Group. (2012). *Statement of Australian Government requirements for environmental information.* Canberra, ACT: Bureau of Meteorology.

Australian National Audit Office. (2008). *Regional delivery model for the natural heritage trust and the national action plan for salinity and water quality.* Canberra, ACT: Departments of the Environment, Water, Heritage and the Arts and Agriculture, Fisheries and Forestry. Retrieved from www.anao.gov.au/uploads/documents/2007-08_Audit _Report_21.pdf

McNaught, I., Thackway, R., Brown, L. & Parsons, M. (2008). *A field manual for surveying and mapping nationally significant weeds* (2nd ed.). Canberra, ACT: Bureau of Rural Sciences.

Miura, S., Amacher, M., Hofer, T., San-Miguel-Ayanz, J., Ernawati & Thackway, R. (2015). Protective functions and ecosystem services of global forests in the past quarter-century. *Forest Ecology and Management 352*, 35–46. doi.org/10.1016/j.foreco.2015.03.039

Montreal Process Implementation Group for Australia and National Forest Inventory Steering Committee. (2013). *State of the forests report.* Canberra, ACT: Australian Bureau of Agricultural and Resource Economics and Sciences.

Mutendeudzi, M. & Stafford-Bell, R. (2011). *Scientific information for making decisions about natural resource management—a report on the value, status and availability of key ABARES datasets* (ABARES Technical Report 11.2). Canberra, ACT: Australian Bureau of Agricultural and Resource Economics and Sciences.

NLWRA (National Land and Water Resources Audit). (2007). *Native vegetation—status of information for reporting against indicators under the national natural resource management monitoring and evaluation framework.* Canberra, ACT: National Land and Water Resources Audit.

NLWRA. (2008a). *Land use—status of information for reporting against indicators under the national natural resource management monitoring and evaluation framework.* Canberra, ACT: National Land and Water Resources Audit.

NLWRA. (2008b). *Land salinity—status of information for reporting against indicators under the national natural resource management monitoring and evaluation framework.* Canberra, ACT: National Land and Water Resources Audit.

Sbrocchi C., Davis R., Grundy M., Harding R., Hillman T., Mount R., … Cosier, P. (2015). *Evaluation of the Australian regional environmental accounts trial.* Sydney, NSW: Wentworth Group of Concerned Scientists.

SoE (State of the Environment Committee). (2011). *Australia state of the environment.* Independent report to the Australian Government Minister for Sustainability, Environment, Water, Population and Communities. Canberra, ACT. Retrieved from www.environment.gov.au/science/soe/2011-report/5-land/2-state-and-trends/2-3-vegetation#ss2-3-1

Thackway, R. (2016). Tracking anthropogenic influences on the condition of plant communities at sites and landscape scales. In A. Z. K. Almusaed (Ed.), *Landscape ecology—the influences of land use and anthropogenic impacts of landscape creation.* London, UK: InTech. doi.org/10.5772/62874

Thackway, R. & Cresswell, I. D. (Eds.). (1995). *An interim biogeographic regionalisation for Australia: A framework for establishing the national system of reserves* (version 4.0). Canberra, ACT: Australian Nature Conservation Agency.

Thackway, R. & Lesslie, R. (2008). Describing and mapping human-induced vegetation change in the Australian landscape. *Environmental Management 42*, 572–90. doi.org/10.1007/s00267-008-9131-5

Thackway, R., Lymburner, L. & Guerschman, J. P. (2013). Dynamic land cover information: Bridging the gap between remote sensing and natural resource management. *Ecology and Society 18*(1), 2. doi.org/10.5751/ES-05229-180102

Thackway, R., Szabo, S. & Smyth, D. (1997). Indigenous protected areas: New opportunities for the conservation of biodiversity. In P. Hale & D. Lamb (Eds.), *Proceedings of the conference on conservation outside of nature reserves* (pp. 62–73). Brisbane, QLD: Centre for Conservation Biology, University of Queensland.

10

Understanding Soil Change: Institutional Requirements to Ensure Australia's National Preparedness

Neil McKenzie

Key Points

- Globally, there is increasing awareness of threats to soil function, finite areas of arable land and apparent yield plateaus of major crops. Better soil management is needed so that nutrients are conserved, water use is improved and emissions are reduced.

- Some of Australia's soil management challenges are immediate and obvious (e.g. widespread acidification of cropping lands). Other problems (e.g. erosion, nutrient imbalances, soil carbon loss and compaction) are subtler, but equally important in the long term.

- A concerted effort to improve soil management in Australia requires improving the diagnostic systems for determining when and where soil function is being compromised; implementing sustainable systems of land use that restore or enhance soil function; and developing more effective institutional arrangements for soil information.

- Over the last 25 years, the provision of soil information in Australia has relied on informal collaboration between national, state and territory agencies. Despite many achievements, the current arrangements are no longer viable.

- Investments into the national soil information system should generate substantial economic benefits. These benefits arise primarily from increases in agricultural productivity and avoidance of costs in other soil-dependent industries. This is before consideration of the equally large societal and ecosystem service benefits associated with better soil and land management.

Introduction

Soils are fundamental to life on Earth but, unlike food, energy, water and air, issues of security, access and quality are less immediate. Despite the patchy and often outdated sources of information on the condition of soil resources globally, there is now sufficient evidence to indicate that threats to soil function require a concerted response. This chapter summarises the observational and analytical capabilities that countries need to have if they are to understand the significance of soil change and manage accordingly. The evolution of Australia's soil information systems is reviewed. It is argued that current collaborative arrangements between state, territory and federal agencies are no longer viable. Proposals for the establishment of a formally mandated agency for mapping, monitoring and forecasting the condition of soils are considered, and it is suggested that integration is necessary with national information systems for land use and management. The resulting land resource information systems are prerequisites for achieving sustainable soil management, mitigating against climate change and optimising the productivity of Australian landscapes (Johnston et al., 2003).

The New Global View on Soil Use and Management

The first State of the World's Soil Resources report by the Intergovernmental Technical Panel on Soils (ITPS, 2015a) concluded that:

> Human pressures on soil resources are reaching critical limits. Further loss of productive soils will amplify food-price volatility and potentially send millions of people into poverty. This loss is avoidable. Careful soil management can increase the food supply, and provides a valuable lever for climate regulation and a pathway for safeguarding ecosystem services. (p. xix)

The ITPS went on to state that:

> While there is cause for optimism in some regions, the overwhelming conclusion from the report is that the majority of the world's soil resources are in only fair, poor or very poor condition. The most significant threats to soil function at the global scale are soil erosion, loss of soil organic carbon, and nutrient imbalance. The current outlook is for the situation to worsen unless concerted actions are taken by individuals, the private sector, governments and international organisations. (p. xix)

Knowledge of soil and land resources is the foundation for achieving sustainable soil management. However, the distribution and characteristics of soils in any district or nation are neither obvious nor easy to monitor. Consequently, understanding whether a land use is well matched to the qualities of the soil requires some form of diagnostic system—both to identify the most appropriate form of management and to monitor how the soil is functioning (McKenzie, 2014). Three important components of the diagnostic system necessary for sustainable land use and management are:

- an understanding of how soils vary across the landscape (e.g. maps of soil properties and functional types)
- an ability to detect and interpret soil change with time (e.g. via monitoring sites and long-term experiments)
- a capacity to forecast the likely state of soils under specified systems of land management and climates (e.g. through the use of simulation models).

The Revised World Soil Charter adopted by all member states of the United Nation's Food and Agriculture Organization (2015) recommended that all nations require coordinated soil information systems similar to those for economic data, weather and water resources that exist in many countries. Further, these national soil information systems need to be integrated with the emerging global soil information system.

The Australian Context

Australia is one of the countries that gave the ITPS some cause for optimism. However, even in Australia, soil acidification, unsustainable rates of soil erosion, loss of soil organic carbon and nutrient imbalances (deficiencies and excesses) are recognised as significant threats to soil function, and remain difficult to ameliorate (ITPS, 2015b). If left unchecked, these problems will constrain Australia's ability to take advantage of agricultural opportunities created by a growing population and demand for exports. A concerted effort to further improve soil management is required; this needs to include not only better diagnostic systems for determining when and where soil function is being compromised, but also effective systems for developing and implementing sustainable systems of land use that restore or enhance soil function. The benefits of achieving sustainable soil management are substantial. They include:

- increased income for farmers and other players within the food supply system
- increased economic activity through the development of service industries that support sustainable soil management
- improved intergenerational equity, particularly for farming families
- more efficient and effective mitigation and adaptation to climate change
- greater food security
- positive externalities including improved water quality and landscape amenity.

It is difficult to estimate the likely return on investments into sustainable soil management. However, the National Committee on Soil and Terrain (NCST, 2013) estimated that an annual investment of $100 million into the national soil information system could generate economic benefits worth $2 billion per annum by 2020. These benefits arise primarily from increases in agricultural productivity and avoidance of costs in other soil-dependent industries that potentially amount to hundreds of millions of dollars per year. This estimate does not include the equally large societal and ecosystem service benefits associated with better soil and land management, particularly carbon sequestration (e.g. Minasny, McBratney, Malone & Stockmann, 2015). These potential benefits are significant for the Australian economy; however, a more thorough analysis is necessary to confirm the scale of returns and to identify priorities for investments.

Institutional Evolution

The evolution of institutions for managing soil resources parallels the recent history of land use in Australia. The initial impact of European colonisation on soils in most parts of Australia was profound; in some areas, catastrophic. The severity of soil degradation, particularly in the 100 years after 1850, was extreme, resulting in declining crop yields and the Dust Bowl years of the 1930s and 1940s (Angus, 2001; Bolton, 1981; McKenzie, Isbell, Jacquier & Brown, 2004; McTainsh & Boughton, 1993). The large economic, social and environmental costs led to a range of institutional responses. At a conference of Commonwealth and state ministers held in Adelaide in August 1936, it was agreed that each state would establish a committee to study the problems of soil erosion and conservation, and suggest solutions; the Council for Scientific and Industrial Research (now the Commonwealth Scientific and Industrial Research Organisation [CSIRO]) was to cooperate with these committees (Australian Bureau of Statistics, 1963; Soil Conservation Committee, 1938). State soil conservation authorities with supporting legislation were subsequently established and coordination was eventually achieved through the Standing Committee of Soil Conservation, which was established in 1946. This committee reported to the Australian Agricultural Council, which had a remit to ensure consultation among Australian governments on economic aspects of primary production. Variants of this arrangement endured for more than 60 years, with the responsibility for soil resources eventually passing to the Natural Resource Management Ministerial Council (NRMMC), which had responsibility for land and water management.

The Collaborative Soil Conservation Study (CSCS, 1978a, 1978b) conducted by the Australian, state and territory governments provided a comprehensive overview of the technical and institutional issues affecting soil management across the country. It laid the groundwork for the establishment of the National Soil Conservation Program in 1983. This coincided with continuing public concern over the extent and severity of land degradation; the large dust storm that engulfed Melbourne in February 1983 brought this into sharp relief. The concern did not abate, but instead contributed directly to the rise of the Landcare movement, which emphasised participative engagement and local action. The Landcare movement began with an unlikely alliance between traditional opponents—conservationists and farmers—and grew into a movement with thousands of groups in Australia and other countries (Campbell, 1992). The activities of Landcare groups transformed many landscapes; large areas were revegetated and restored through their efforts. The strength of Landcare lay in the community groups and networks that conceived their own visions and set goals for local and regional environmental action, with government and corporate support (Youl, Marriot & Nabben, 2006).

Unprecedented investment into natural resource management from the Australian Government was associated with the rise of Landcare. Several large natural resource management programs built upon the National Soil Conservation Program—the National Landcare Program (1989), Natural Heritage Trust (1997) and Caring for our Country (2008) invested billions of dollars into natural resource management. It is difficult to accurately assess the effects of these programs, for although significant improvements in natural resource conditions have been achieved, it is generally acknowledged that the scale of management actions may only be slowing, rather than reversing, the negative impacts that would occur without intervention (e.g. Auditor General, 2008; Australian Government, 2013; Pannell et al., 2012).

Major land use conflicts over the use of high-quality soils have figured prominently during the last decade (e.g. Williams, 2015); however, the scale of investment into general natural resource management programs has declined. The reasons are complex, but the following factors are significant:

- major improvements in soil and land management during the last 25 years (e.g. ITPS, 2015b; State of the Environment [SoE], 2011), which have quite likely contributed to a perception that soil and land management problems have been solved

- the difficulty of demonstrating returns on investment from large natural resource management programs, resulting in public funds being directed elsewhere (e.g. health, education and national security)

- the concerns over some major problems such as dryland salinity lessening, partly due to seasonal climate shifts and also because the initial projections were overstated

- the end of the Millennium Drought (van Dijk et al., 2013), water reforms and significant improvements in water resource information systems contributing to a sense that water problems have been solved

- the Global Financial Crisis and end of the resources boom forcing governments to reduce expenditure

- government agencies responsible for natural resource management struggling to find a compelling narrative and mode of operation that could compete with other government priorities—one indication of this was the abolition of the NRMMC and Land and Water Australia.

Despite these developments, many of the less obvious—but chronic— issues affecting the soil resources of Australia remain. However, it is unreasonable to expect governments and industries to invest in sustainable soil management unless there is compelling evidence that these chronic issues require a response. Such evidence will not be forthcoming until there is an overhaul of Australia's soil information system.

Australia's Soil Information System

Most reviews of the soil knowledge system in Australia highlight institutional complexity, inconsistency of technical methods, limited economies of scale, ineffective mechanisms for funding and lack of a long-term strategy (e.g. Beckett & Bie, 1978; Campbell, 2006; CSCS, 1978a, 1978b; McKenzie, 1991; Taylor, 1970; Wood & Auricht, 2011). While some of these problems have been solved, significant institutional constraints remain. McKenzie (2014) summarised these constraints as follows:

- All levels of government need reliable information on soil resources, but no single level of government or department has responsibility for collecting this information on behalf of other public sector agencies.

- Public and private interests in soil are large and overlapping, but mechanisms for co-investment by public and private agencies have not been developed.
- Market failure in relation to the supply and demand of soil information is a significant and widespread problem. In the simplest case, beneficiaries of soil information do not pay for its collection and this reduces the pool of investment for new survey, monitoring and experimental programs.
- Partly, as a result of the above, most soil information–gathering activities are currently funded through short-term government programs, private companies and individuals, or in response to specific regulatory requirements (e.g. environmental impact statements). These have not produced the enduring, accessible and broadly applicable information systems that are needed to meet the requirements of nearly all stakeholders.

Despite these significant challenges, the Australian soil information system is recognised for being innovative, collaborative and responsive to contemporary issues. This is largely due to the enduring and effective partnerships between operational agencies and research groups that have been responsible for a range of innovations, including digital soil mapping, proximal sensing and web-based delivery of information services (Arrouays et al., 2014a, 2014b; Grundy et al., 2015; Hicks, Rossel & Tuomi, 2015).

The Collaborative Model

Despite several previous proposals for the establishment of a national soil information agency (CSCS, 1978b; Taylor, 1970), a strategic review of soil survey and land evaluation activities by McKenzie (1991) concluded that a voluntary and collaborative model was most appropriate for addressing the technical and institutional problems apparent at that time. As a result, the Australian Collaborative Land Evaluation Program (ACLEP) was established to develop a coordinated approach to land resource assessment across Australia (Hallsworth, 1978). The program included all Australian, state and territory government agencies involved with land resource assessment. ACLEP was jointly funded by the Australian Government (initially through its National Landcare Program) and CSIRO. In many ways, the model was a continuation of the original institutional arrangements established in 1936.

ACLEP aimed to encourage sustainable land use and environmental protection in Australia by promoting better procedures for acquiring and using land resource information in government and private industry. This was achieved by setting national standards for land resource assessment, providing a forum for communication between technical specialists, attempting to develop a network of soil and land reference sites across Australia and encouraging research into methods for land resource assessment. ACLEP received strategic direction from the NCST (formerly the Working Group on Land Resource Assessment) and, for most of its existence, has had a formal line of reporting to the relevant ministerial council.[1] In its current form, ACLEP has the following objectives:

- provide coordinated, scientific assessment and monitoring of Australia's soil and land resources
- develop and promote methods and standards for soil measurement, land resource survey and monitoring of soil condition
- build and support coordination, collaboration, partnerships and development of skills and capacity through improved communication
- be the primary contact point for information on Australia's soil and land resources and related national assessments
- provide a focus for the collection, collation, management, dissemination and analysis of nationally consistent, integrated data and information on soil and land resources through the Australian Soil Resource Information System and the CSIRO National Soil Archive.

During its 23 years of operation, ACLEP has relied primarily on project funding from the Australian Government (the Department of Agriculture and Water Resources and its precursors) with matching resources from CSIRO. It has also been able to entrain broader support from CSIRO (e.g. computing infrastructure) and in-kind contributions from state and territory agencies.

ACLEP has had three main phases:

1. 1990–2000: this phase coincided with the Decade of Landcare when state and territory agencies, with partnership funding from the Australian Government, undertook the Accelerated Program of Land

---

1   Initially, the ACLEP reported to the Australian Soil Conservation Council, later to the Natural Resources Ministerial Council, and then to the Primary Industries Ministerial Council until it was disbanded in 2013. It now reports indirectly to the Standing Council on Primary Industries, but the scope of the NCST extends well beyond agriculture.

Resource Assessment. Some states (e.g. South Australia and Western Australia) were able to complete new surveys across their agricultural lands. Other states and territories adopted different strategies. During this period, ACLEP focused on supporting the partner agencies through capacity building, development and publication of standards, and testing of new survey methods. The ACLEP newsletter provides a comprehensive catalogue of activities during this period.

2. 2000–10: the focus during this period was on the communication of soil and land resource information to a broad range of users, while at the same time providing technical support for new methods of digital soil mapping (e.g. Henderson, Bui, Moran & Simon, 2005; McKenzie, Grundy, Webster & Ringrose-Voase, 2008). The national soil information system was upgraded to become one of the world's first online national soil information systems. Collaborations were pursued with other technical groups to provide assessments and information at the continental scale (e.g. NLWRA, 2001, 2002; Peverill, Sparrow & Reuter, 1999). ACLEP was also heavily involved in developing technical recommendations and guidance on monitoring the condition of Australian soils (Grealish, Clifford, Wilson & Ringrose-Voase, 2011; McKenzie & Dixon, 2007; McKenzie, Henderson & McDonald, 2002). This period saw initial steps taken to build stronger links between technical programs and proposals for a national soil policy (Campbell, 2008). A significant effort was directed towards improving public understandings of soils and soil management (McKenzie et al., 2004).

3. 2010–15: with the exception of Tasmania and the Northern Territory, most field survey and monitoring programs had been curtailed by this time, and the focus for ACLEP was on improving online access to existing soil information. Opportunities in geospatial technologies led to the development of new standards for soil data systems and web-based services. Ongoing support was provided for synoptic assessments (e.g. ITPS, 2015b; SoE, 2011), and a major upgrade of the National Soil Archive was completed (Karssies & Wilson, 2015). The latter activity was especially significant because it allowed new methods of proximal sensing to be deployed on archived samples and, in the process, generated large new datasets for the country (e.g. Hicks et al., 2015). The most significant achievement of this period involving all partners was the development of the Soil and Landscape Grid of Australia (Grundy et al., 2015). This was an internationally significant achievement because it was the first continental-scale implementation

of the GlobalSoilMap technical specifications (Arrouays et al., 2014a) that are a key component of the emerging global soil information system (Global Soil Partnership, 2014).

It is difficult to provide a detached and objective view of the effectiveness of the collaborative model, and ACLEP in particular. However, it is worth noting the following:

- Users of soil information now have unprecedented access to harmonised soil data and information collected over more than 50 years.
- Major technical advances and major new products have been delivered because of the network and collaborative arrangements fostered by ACLEP.
- ACLEP provided a particularly effective pathway-to-impact for research teams in universities (e.g. University of Sydney) and CSIRO.

Despite the long record of achievement, it would appear that the collaborative model forged by ACLEP is no longer viable for several reasons:

- Most state agencies have stopped their field programs of soil survey and monitoring. As a result, the map coverage is now out-of-date and monitoring networks are not being established or maintained.
- A closely related issue is the demographic profile of the current cohort of experienced pedologists. Most were trained in the 1980s and 1990s and participated in the Accelerated Program of Land Resource Assessment. The expertise, and especially the field knowledge, held by these experts is not being passed onto a new generation of soil and land resource specialists.
- The formal programs for funding research (e.g. via the Rural Research and Development Corporations and the Australian Research Council) are more enduring and better organised than those for operational programs of land resource survey and monitoring. At present, there are very few sources of funds for the latter.
- ACLEP and its related activities have been funded through the major natural resource management programs listed earlier. However, none of these programs are compelled to fund operational survey and monitoring programs, and there are no jurisdictional mandates to compel governments to continue such programs. This contrasts with

other areas, such as weather and climate, in which legislation provides the formal basis for data-gathering programs. The levels of funding for ACLEP have declined in real terms in recent years.

- There has been a general trend towards smaller national, state and territory agencies—the traditional homes for land resource survey and monitoring.

- The CSIRO has maintained its long tradition of supporting land resource surveys, but the investment is considerably less than in previous decades when several divisions were actively involved (e.g. the former Division of Soils, and the Division of Land Use Research and its successors). The CSIRO continues to have a role in undertaking research and development to support survey and monitoring programs; however, it has no formal mandate to provide resources for the ongoing operational activities that are necessary if Australia is to have the information services it needs to ensure sustainable soil management.

Barriers to Putting Soils on the National Agenda

Institutional reform is necessary if Australia is to have a national soil information system that is compatible with the broader global effort. Any significant institutional reform requires public support and engagement by policymakers, politicians, industry groups and civil society. However, numerous significant barriers limit public awareness of soil issues— for example:

- Most people do not have a clear view of the condition of the soil resources upon which their lives ultimately depend. One cause is increasing urbanisation and the reality that the proportion of human labour devoted to working the soil has steadily decreased through the past century. Further, most people are now protected from local resource depletion due to trade and the area of land and water used to support them being scattered all over the planet (ITPS, 2015a).

- Most threats to soil function are chronic and long-term—soil erosion, acidification and depletion of carbon and nutrients occur over decades. Some of these changes can be difficult to detect and there is a risk that management responses will not occur until critical and irreversible thresholds have been exceeded.

- Institutional responses to natural resource problems are often triggered by major polarising events (e.g. droughts, fires and floods). Apart from dust storms, which are much less frequent than during the 1930s and 1940s (SoE, 2011), changes in soil condition rarely rate a mention in the mainstream news media, although soil contamination is an occasional exception because it may directly affect human health and food quality.

- For most primary industries (e.g. cropping, grazing, forestry and horticulture), soils are a means to an end. Management systems tend to focus on the final product or readily measured indicators along the supply chain (e.g. yield or market price). Soils are acknowledged as important; however, other factors that have an immediate impact on profitability are often prioritised. Consequently, insufficient investment into research, development and extension can easily occur.

- Soil scientists who draw attention to the lack of investment into soil activities can be readily dismissed as self-interested. The recent emergence of independent public advocates on soil issues has started to address this problem (e.g. former diplomat and Governor of Queensland Penny Wensley and Australia's Advocate for Soil Health Major General Michael Jeffery).

- There have been major improvements in soil and land management during the last 25 years—conservation farming, controlled traffic, cell grazing and the more general achievements of the Landcare movement (e.g. Natural Decisions Pty Ltd, 2015). This has likely contributed to a perception that soil and land management problems have been solved.

- Australia is a food exporter and its citizens have access to an abundance of inexpensive, high-quality food. Issues relating to food security do not figure in public discourse, except in relation to international affairs, and when famines occur in distant countries.

Additional factors limiting the consideration of soils by policymakers include a lack of ready access to the evidence needed for policy action and the challenge of dealing with property rights for a natural resource that is often privately owned, yet vital for the public good (ITPS, 2015a). All these factors contribute to our current lack of institutional preparedness—both domestically and internationally.

The Public Sector Role

The decision, in 1991, to establish ACLEP using a voluntary and collaborative model was appropriate, particularly given the technical strength of some state and territory agencies. However, the situation has changed. A more enduring and self-sustaining system is now required, not only for soils, but also for natural resource information more generally.

Craemer and Barber (2007) outlined some of the market failure arguments relevant to the development of a 'business case' for public investment in soil information. These related to the presence of externalities, which were important to basic research and development, and information failures within markets. They noted the significance of information as infrastructure, and the importance of soil information being collected once and then used for many different purposes. While Craemer and Barber provided a valuable starting point for developing the economic case for investing in soil information, more work still needs to be done. It is assumed that there is a legitimate role for the public sector in gathering and providing soil information. However, given the private- and public-good nature of soil resources, it is also assumed that some form of public–private model would be appropriate.

The NCST (2013) has recently proposed a comprehensive plan for re-engineering the national soil information infrastructure, so that it can provide the required data and information to regularly assess the condition of soils and their responses to land management across Australia. A prerequisite for implementing the plan is establishing a formal and enduring mandate for soil resource assessment. One possible mechanism is via the incorporation, by legislation, of soil resource assessment activities into one or more agencies. This mandate is essential because of the long time frames required to build the soil resource information base and monitor soil change over several decades. The second key institutional issue identified by the NCST relates to the organisational and business model. Three options were proposed:

1. Create a new Bureau of Soil Resources modelled on the much larger Bureau of Meteorology and Australian Bureau of Statistics: the Bureau of Soil Resources would have a legislated charter to serve all levels of government and engage with private sector activities to maximise the net benefit for Australia. The bureau would be responsible for survey, monitoring and technical activities. It would have a workforce

of technical specialists and significant capital assets (e.g. field survey capability, laboratories and computing infrastructure). The Bureau of Soil Resources could be administered by an existing agency to minimise administrative costs.

2. Establish a legislated program within an existing Australian government agency: the program would have a central management team to coordinate the survey, monitoring and technical activities, but most of these would be contracted out to either state and territory agencies or the private sector. The balance between in-house technical work and externally contracted work would require careful management to avoid the loss of corporate knowledge and intellectual capital that can occur with such arrangements.

3. Formation of a new organisation with a business model and structure similar to current arrangements for Cooperative Research Centres: the legislation and agreements supporting such a centre would differ from that of a Cooperative Research Centre because of the requirement for the organisation to be enduring.

The experience of the last 15 years suggests that the first option is preferable, although it is recognised that Australian governments have little appetite for establishing new organisations (e.g. Morton & Tinney, 2012).

Soil Resources and Land Management

The proposal by the NCST to establish an Australian Soil Assessment Program has recently stalled, primarily because of an institutional impasse and lack of political momentum. This may change; governments may see land use policy as once again being central to national prosperity. Hatfield-Dodds et al. (2015) outlined how new technologies and incentives could decouple economic and environment outcomes to enable progress towards sustainable prosperity. However, this requires greater commitment to land use planning and management, along with the necessary information systems as support.

If the activities of the land use mapping community were integrated with those traditionally undertaken by ACLEP, a strong case could be mounted for investment in natural resource information systems. Given that the Australian Collaborative Land Use and Management Program (ACLUMP) was founded on the same principles as ACLEP, and that many of the issues facing the land use mapping community are similar to those outlined above for the soil resource community, this is a natural development. The late Rob Lesslie was a leader in the establishment and operation of ACLUMP; he was extremely keen to integrate the work of his community with the work of ACLEP. The establishment of an Australian land management agency that would provide technical and policy support to achieve the sustainable and prosperous future identified by Hatfield-Dodds et al. (2015) would be a fitting memorial.

References

Angus, J. F. (2001). Nitrogen supply and demand in Australian agriculture. *Animal Production Science 41*, 277–88. doi.org/10.1071/EA00141

Arrouays, D., Grundy, M. G., Hartemink, A. E., Hempel, J. W., Heuvelink, G. B. M., Hong, S. Y., … Zhang, G. L. (2014a). GlobalSoilMap: Toward a fine-resolution global grid of soil properties. *Advances in Agronomy 125*, 93–134. doi.org/10.1016/B978-0-12-800137-0.00003-0

Arrouays, D., McKenzie, N. J., Hempel, J., Richer de Forges, A. C. & McBratney, A. B. (Eds.). (2014b). *GlobalSoilMap—basis of the global spatial soil information system: Proceedings of the 1st GlobalSoilMap Conference*. London, UK: Taylor & Francis Group.

Auditor General. (2008). *Regional delivery model for the Natural Heritage Trust and the national action plan for salinity and water quality. Australian National Audit Office* (Audit Report No. 21). Canberra, ACT. Retrieved from www.anao.gov.au/sites/g/files/net3721/f/ANAO Report_2007-2008_21.pdf

Australian Bureau of Statistics. (1963). *Soil conservation* [Year Book Australia 1963]. Retrieved from www.abs.gov.au/AUSSTATS/abs@.nsf/allprimarymainfeatures/816FB6F65D2BA89ECA2573AE00045D33?opendocument

Australian Government. (2013). *Caring for our country 2008–2013. Achievements report: The synthesis chapter*. Retrieved from www.nrm.gov.au/publications/achievements-report-synthesis

Beckett, P. H. T. & Bie, S. W. (1978). *Use of soil and land system maps to provide soil information in Australia.* Melbourne, VIC: CSIRO.

Bolton, G. C. (1981). *Spoils and spoilers: Australians make their environment 1788–1980.* Sydney, NSW: Allen & Unwin.

Campbell, A. (1992). *Landcare in Australia.* Canberra, ACT: National Soil Conservation Program.

Campbell A. (2006). *The Australian natural resource management knowledge system.* Canberra, ACT: Land and Water Australia.

Campbell, A. (2008). *Managing Australia's soils. A policy discussion paper.* Retrieved from www.clw.csiro.au/aclep/documents/Soil-Discussion-Paper.pdf

CSCS (Collaborative Soil Conservation Study). (1978a). *A basis for soil conservation policy in Australia. Commonwealth and state government Collaborative Soil Conservation Study 1975–77* (Report 1). Canberra, ACT: Australian Government Publishing Service.

CSCS. (1978b). *Towards a national approach to land resource appraisal. Commonwealth and state government collaborative soil conservation study, 1975–77* (Report 2). Canberra, ACT: Australian Government Publishing Service.

Craemer, R. & Barber, M. (2007). *Building a business case for investment in soil information.* Canberra, ACT: National Land and Water Audit and National Heritage Trust.

Food and Agriculture Organization. (2015). *Revised world soil charter.* Retrieved from www.fao.org/3/a-mn442e.pdf

Global Soil Partnership. (2014). *Plan of action for pillar four of the Global Soil Partnership adopted by the GSP plenary assembly.* Retrieved from www.fao.org/3/a-az921e.pdf

Grealish, G., Clifford, D., Wilson, P. & Ringrose-Voase, A. (2011). *National soil condition monitoring for soil pH and soil carbon: Objectives, design, protocols, governance and reporting* (Report 05/11 for Caring for our Country). Canberra, ACT: CSIRO Land and Water Science.

Grundy, M. J., Rossel, R. V., Searle, R. D., Wilson, P. L., Chen, C. & Gregory, L. J. (2015). Soil and landscape grid of Australia. *Soil Research 53*, 835–44. doi.org/10.1071/sr15191

Hallsworth, E. G. (1978). *Purposes and requirements of land resource survey and evaluation. Commonwealth and state government collaborative soil conservation study* (Report 3). Canberra, ACT: Department of Environment, Housing and Community Development, AGPS.

Hatfield-Dodds, S., Schandl, H., Adams, P. D., Baynes, T. M., Brinsmead, T. S., Bryan, B. A., … McCallum, R. (2015). Australia is 'free to choose' economic growth and falling environmental pressures. *Nature 527*, 49–53. doi.org/10.1038/nature16065

Henderson, B. L., Bui, E. N., Moran, C. J. & Simon, D. A. P. (2005). Australia-wide predictions of soil properties using decision trees. *Geoderma 124*, 383–98. doi.org/10.1016/j.geoderma.2004.06.007

Hicks, W., Rossel, R. V. & Tuomi, S. (2015). Developing the Australian mid-infrared spectroscopic database using data from the Australian soil resource information system. *Soil Research 53*, 922–31. doi.org/10.1071/sr15171

ITPS (Intergovernmental Technical Panel on Soils). (2015a). *Status of the world's soil resources (SWSR)—technical summary* [Food and Agriculture Organization of the United Nations' website]. Retrieved from www.fao.org/documents/card/en/c/39bc9f2b-7493-4ab6-b024-feeaf49d4d01/

ITPS. (2015b). Regional assessment of soil change in the southwest Pacific. In *SWSR Report* (Chapter 15) [Food and Agriculture Organization of the United Nations' website]. Retrieved from www.fao.org/documents/card/en/c/39bc9f2b-7493-4ab6-b024-feeaf49d4d01/

Johnston, R. M., Barry, S. J., Bleys, E., Bui, E. N., Moran, C. J., Simon, D. A. P. … Grundy, M. (2003). ASRIS: The database. *Australian Journal of Soil Research 41*, 1021–36. doi.org/10.1071/SR02033

Karssies, L. & Wilson, P. (2015). Soil archives: Supporting research into soil changes. *IOP Conference Series: Earth and Environmental Science 25*(1). doi.org/10.1088/1755-1315/25/1/012021

McKenzie N. J. (1991). *A strategy for coordinating soil survey and land evaluation in Australia.* Glen Osmond, SA: CSIRO, Division of Soils.

McKenzie, N. J. (2014). Australian soils. In D. Lindenmayer, S. Dovers & S. Morton (Eds.), *Ten commitments revisited.* Melbourne, VIC: CSIRO Publishing.

McKenzie, N. J. & Dixon, J. C. (Eds.). (2007). *Monitoring soil condition across Australia: Recommendations from the expert panels.* Retrieved from www.asris.csiro.au/downloads/Monitoring%20Soil%20Condition %20ver_4.doc

McKenzie, N. J., Grundy, M. J., Webster, R. & Ringrose-Voase, A. J. (Eds.). (2008). *Guidelines for surveying soil and land resources* (vol. 2). Melbourne, VIC: CSIRO Publishing.

McKenzie, N. J., Henderson, B. & McDonald, W. S. (2002). *Monitoring soil change. Principles and practices for Australian conditions.* Canberra, ACT: CSIRO Land and Water.

McKenzie, N. J., Isbell, R. F., Jacquier, D. W. & Brown, K. L. (2004). *Australian soils and landscapes: An illustrated compendium.* Melbourne, VIC: CSIRO Publishing.

McTainsh, G. & Boughton, W. C. (Eds.). (1993). *Land degradation processes in Australia.* Melbourne, VIC: Longman Cheshire.

Minasny, B., McBratney, A. B., Malone, B. & Stockmann, U. (2015, 1 December). Eyes down: How setting our sights on soil could help save the climate. *The Conversation.* Retrieved from theconversation. com/eyes-down-how-setting-our-sights-on-soil-could-help-save-the-climate-51514

Morton, S. & Tinney, A. (2012). *Independent review of Australian government environmental information activity: Final report.* Canberra, ACT: Australian Government. Retrieved from www.environment.gov. au/resource/independent-review-australian-government-environmental -information-activity

NCST (National Committee on Soil and Terrain). (2013). *Establishing the Australian cooperative soil assessment program. Supporting Australia's sustainable future through improved knowledge of Australian soils and their responses to land management.* A report prepared by the National Committee on Soil and Terrain for the Soil Research Development and Extension Reference Group.

Natural Decisions Pty Ltd. (2015). *Evidence for the economic impacts of investment in national Landcare programme activities* [Australian Government website]. Retrieved from www.nrm.gov.au/publications/economic-impacts-nlp-activities

NLWRA (National Land and Water Resources Audit). (2001). *Australian agricultural assessment 2001.* Canberra, ACT: National Land and Water Resources Audit.

NLWRA. (2002). *Australia's natural resources: 1997–2002 and beyond.* Canberra, ACT: National Land and Water Resources Audit.

Pannell, D. J., Roberts, A. M., Park, G., Alexander, J., Curatolo, A. & Marsh, S. P. (2012). Integrated assessment of public investment in land-use change to protect environmental assets in Australia. *Land Use Policy, 29*, 377–87. doi.org/10.1016/j.landusepol.2011.08.002

Peverill, K. I., Sparrow, L. A. & Reuter, D. J. (1999). *Soil analysis: An interpretation manual.* Melbourne, VIC: CSIRO Publishing.

Soil Conservation Committee. (1938). *Report of the soil conservation committee.* Adelaide, SA: Government Printer. Retrieved from history.pir.sa.gov.au/__data/assets/pdf_file/0011/168266/Soil_Conservation2.pdf

SoE (State of the Environment Committee). (2011). *Australia state of the environment 2011: An independent report to the Australian government minister for sustainability, environment, water, population and communities.* Canberra, ACT: Department of Sustainability, the Environment, Water, Population and Communities.

Taylor, J. K. (1970). *The development of soil survey and field pedology in Australia, 1927–67.* Canberra, ACT: CSIRO Australia.

van Dijk, A. I .J. M., Beck, H. E., Crosbie, R. S., de Jeu, R. A. M., Liu Y. Y., Podger, G. M., … Viney, N. R. (2013). The millennium drought in Southeast Australia (2001–2009): Natural and human causes and implications for water resources, ecosystems, economy and society. *Water Resources Research 49,* 1–18. doi.org/10.1002/wrcr.20123

Williams, J. (2015). Soils governance in Australia: Challenges of cooperative federalism. *International Journal of Rural Law and Policy 1*, 4173.

Wood B. G. & Auricht C. M. (2011). *ASRIS/ACLEP user needs analysis.* Brighton, SA: Auricht Projects.

Youl, R., Marriot, S. & Nabben, T. (2006). *Landcare in Australia: Founded on local action.* Wallington, VIC: SILC and Rob Youl Consulting Pty Ltd.

Part 4 – Social and Natural Drivers of Change

11

Environmental Conflict: Engaging with Scientific Information and Community Activism

Jacki Schirmer

Key Points

- Land and water use planning often involves managing environmental conflict.

- Environmental conflicts are complex and can involve a wide range of substantive, procedural and psychological issues. They share one key characteristic: scientific evidence often plays a key role.

- Successful management of environmental conflict requires carefully considering how to ensure that scientific evidence is brought to bear in a way that supports dialogue, rather than deepening divisions and differences of view.

- This can be achieved by ensuring that agreement is reached between the parties involved in environmental conflict on the shared values that underpin the interpretation of scientific evidence, what is considered to be good-quality science (and the thresholds used to assess this) and how future 'unknowns' (in the form of new and emerging issues) will be dealt with to reduce the likelihood of conflict continuing to re-occur over time.

Box 11.1: Examples of the role of information in facilitating dialogue in environmental conflicts

Environmental conflict	Background	How scientific information was used to try to resolve this conflict
Regional Forest Agreements	The RFA process was a response to decades of conflict over the harvesting of timber from native forests in Australia. In several regions across Australia, multi-stakeholder groups were convened from the mid-1990s to the early 2000s to engage in dialogue and agree on areas of forest to be reserved and those to be made available for harvest into the future. The goal was to sign RFAs between state and federal governments that guaranteed wood supply for 20 years, with the agreements reviewed every five years. RFAs were signed in most regions, except for Queensland. However, conflict over timber harvesting has continued in most regions since the RFAs were signed (Musselwhite & Herath, 2005).	The RFAs included a strong focus on gathering and assessing scientific evidence to inform discussions. A group of scientific experts developed the 'JANIS criteria' (JANIS, 1997)—a set of standards for the reservation of different forest types that specified minimum proportions of different forest types to be placed in reserves, among other things. However, critics complained that the JANIS criteria were 'watered-down' by government representatives after their initial formulation (Kirkpatrick, 1998). A series of Comprehensive Regional Assessments were carried out to collate existing information and, in some cases, collect new data. These data informed discussions and decision-making. However, the Comprehensive Regional Assessment process, which sometimes failed to integrate different perspectives and values, was criticised for being rushed (Brueckner & Horwitz, 2005). While RFAs include a five-yearly review process, which requires monitoring of the implementation of the RFA, there has not been ongoing stakeholder dialogue regarding the scientific evidence for achieving desired RFA outcomes.

Environmental conflict	Background	How scientific information was used to try to resolve this conflict
Murray–Darling Basin Plan	The Murray–Darling Basin Plan was a response to concerns about the over-allocation of water within the Murray–Darling Basin, and a desire to provide more sustainable irrigation water supplies together with greater water delivery to many important wetland, river and nature areas in the Basin. The plan was developed by the Murray–Darling Basin Authority (MDBA). The development process included initial consultations, after which a guide to the proposed plan was produced. This was partly in response to concerns about the relatively centralised process of the plan's development to that point (Daniell, 2011). The guide, which set out options for the plan, was highly controversial; its release was met with protests in several irrigation-dependent communities (Quiggin, 2012). The MDBA subsequently engaged in further consultation with a wide range of groups before producing the final plan, which was legislated by the Australian Government in November 2012.	Multiple expert scientific assessments were commissioned by the MDBA to inform development of the Murray–Darling Basin Plan (Young, Bond, Brookes, Gawne & Jones, 2011) and an advisory committee was established to provide expert advice on how scientific knowledge should be used in establishing the plan (MDBA, 2014). However, many stakeholder groups mistrusted these assessments (Daniell, 2011; MDBA, 2014). This led to the commissioning of alternative assessments by some stakeholder groups—some of which offered widely differing conclusions (Regional Development Australia Northern Inland, n.d.) and analyses (e.g. Grafton & Jiang, 2011). The Murray–Darling Basin Plan includes monitoring and evaluation requirements over time; however, the level of resourcing for the assessment of scientific data as part of these processes is unclear.

Environmental conflict	Background	How scientific information was used to try to resolve this conflict
Tasmanian forest peace process	The Tasmanian forest peace process was a three-year process of negotiations between stakeholders involved in conflicts over the harvesting of timber from native forests in Tasmania. Representatives from the forest industry, environmental groups, unions and community groups met to try to agree on which area of the native forest estate in Tasmania currently available for timber harvesting would be placed in reserves. Government representatives observed the negotiation process in later stages. The process resulted in an agreement that was accepted by the Tasmanian Government and legislated as the Tasmanian Forest Agreement (TFA) in April 2013. However, after state and federal elections returned conservative governments, the TFA was repealed in 2014 (Schirmer, Dare & Ercan, 2016).	Scientific evidence was central to the discussions. After a year of negotiations in which differences in interpretation of available evidence by different parties became apparent, a set of principles for further steps to achieve a peace agreement was created. These steps included a process for assessing the available scientific evidence to assist in answering key questions that were central to the negotiations; the questions focused on the environmental values of forests, sustainable wood supplies and the socio-economic effects of the changes being proposed. An IVG was established in 2011. This group of experts was asked to assess the evidence and report to the negotiating parties in a relatively short time frame (West, 2012). Although not uncontested (Poynter, 2013), the IVG's report informed subsequent negotiations; however, some felt it had limited influence (Schirmer et al., 2016).

Introduction

Land use planning in Australia often involves the management of environmental conflict: disagreements over how best to manage Australia's land and water resources are common and have been for many years (Mercer, 2000). Environmental conflicts are diverse, ranging from disputes about local land use developments, such as new wind farms or intensive livestock farming operations, to large-scale protests and concerns about issues such as the harvesting of timber from native forests, coal seam gas extraction or water allocation in the Murray–Darling Basin, to provide just a few examples.

Like any form of social conflict, environmental conflict is driven by many factors, including substantive, procedural and psychological interests. Substantive issues focus on the claims of fact being made, such as differing views about the environmental impacts of a particular land use. Procedural issues focus on the processes by which people raise concerns and engage in dialogue about environmental issues. Psychological interests refer to the relationships between people involved in the conflict—they have an important effect on the likelihood of successfully addressing the conflict (Furlong, 2010). Scientific information and evidence plays an important role in environmental conflict. It is frequently utilised by parties engaged on different sides of a conflict to support their points of view or provide 'definitive' answers. This sometimes results in 'science wars', in which different parties put forward differing claims about the scientific evidence for an environmental issue (Harding, Hendriks & Faruqi, 2009; Schirmer, 2013; Wynne, 2006).

Scientific evidence has much asked of it in environmental conflicts; the hope that it can be used to resolve conflicts by providing unbiased, independent evidence is often misplaced. Scientific evidence, like any form of social knowledge, reflects the conscious and unconscious values and positions held by those who produce it and interpret its meaning. Frequently funded by those with particular points of view, the scientific evidence produced for environmental conflicts may focus on answering questions that are constructed in ways that favour one side or another, and may be interpreted in different ways depending on values and points of view. Consequently, scientific data can as readily exacerbate conflict as resolve it (Schirmer, 2013).

Yet, despite these challenges, scientific evidence has a critical role to play in environmental conflicts. The solution to the problems that arise when attempting to draw on this evidence is not to attempt to reach 'value-free' science; instead, those involved in dialogue about environmental issues should actively discuss and agree upon the values and perspectives underpinning the collection and interpretation of scientific data (Longino, 1990). Scientific evidence can only contribute constructively to environmental conflict resolution if careful attention is paid to how, when and why the evidence was collected, used and interpreted.

How, then, can scientific data be used constructively, rather than destructively, to help address environmental conflict in Australia? This chapter proposes three principles, based on modern conflict resolution theory, for the constructive use of scientific data: agreeing on values, agreeing on how to 'do the science' and agreeing on how to deal with the 'unknowns'. These are examined in the context of a review of the scientific data used in attempts to resolve three prominent Australian environmental conflicts: Regional Forest Agreements (RFAs), the Murray–Darling Basin Plan and the Tasmanian forest peace process (see Box 11.1).

Principle 1: Agree on Your Values

Science is not value free. Ideally, the values that underpin the interpretation of scientific evidence should be agreed upon by the different groups involved in a conflict *prior* to assessment of the scientific evidence. Enabling agreement to be reached on shared values and interpretation techniques will help to ensure that scientific evidence supports productive dialogue, rather than further enhancing disagreement between stakeholders.

To some extent, the RFA process achieved this by implementing a set of criteria—known as the 'JANIS criteria'—that specified the proportion of forests of different types that should be reserved to create a comprehensive, adequate and representative reserve system. The criteria were agreed upon by the various parties involved in the negotiations. While the final JANIS criteria were criticised by some scientists and environmental groups, they were nevertheless useful, as they provided a set of values to guide decision-making based on the scientific evidence examined in the RFA negotiations. In the Tasmanian forest peace process, stakeholders reached early agreement on the principles by which their negotiations would proceed, including, for example, agreement that the negotiating parties would seek to support a sustainable native forest industry; later, this assisted in the use and interpretation of scientific data. By contrast, in the Murray–Darling Basin Plan process, there was no agreement on values between stakeholders. When there was protest regarding the extent to which socio-economic effects should be considered in determining sustainable levels of water diversion in the Basin, this lack of agreement on values exacerbated the conflict.

It is not easy to create space to discuss values prior to assessing scientific evidence in environmental conflicts; however, doing so can assist in building consensus on the interpretation of available scientific evidence. The key areas to be discussed are the subjective values and thresholds to be applied when making decisions based on science, such as the area of a representative ecotype that should be preserved, or the weight to be given to social and economic effects versus ecological outcomes of a land use decision.

Principle 2: Agree How You Will Do the Science

Scientific evidence is only effective if all stakeholders trust the way the evidence was collected, the people who collected the evidence and the way data have been interpreted. In an ideal world, these things would be agreed upon prior to the collection of scientific data. In reality, it is usually not possible to collect large volumes of new scientific evidence to inform stakeholder discussions in environmental conflicts. Instead, existing scientific evidence must be relied upon—that is, evidence collected for a wide range of purposes by a wide range of researchers. Agreeing how to interpret this evidence and assess its validity and reliability is an essential first step to enabling constructive use of scientific evidence as part of dialogue. Ensuring that people with different viewpoints and perspectives are involved in reaching this agreement contributes to improved procedural justice, which in turn assists in achieving constructive dialogue between parties. Agreeing on the principles by which the quality, comprehensiveness and adequacy of scientific evidence will be evaluated assists in addressing substantive issues of the conflict, as it reduces the likelihood that the same evidence will be interpreted differently by different parties. Therefore, ensuring a shared approach to understanding the science of the conflict enables better utilisation of scientific evidence to address the substantive issues at the heart of many environmental conflicts.

This process of agreement on how scientific evidence will be used, interpreted and understood prior to its evaluation is not commonly undertaken, and the extent to which it occurred in the three conflicts examined in this chapter is not always clear. In the RFAs and the Murray–Darling Basin Plan, a lack of agreement on the meaning and interpretation of the available science was apparent both before and after attempts at negotiating the conflicts. The Tasmanian peace process was somewhat different. All the negotiating parties explicitly agreed to the appointment of the experts who assessed the available scientific evidence—that is, the chosen experts were trusted in theory by all parties to assess and interpret available data. In addition, the negotiating parties explicitly agreed on the type of data to be evaluated and the questions to be answered in the evaluation of data. This agreement on the parameters of the assessment helped to ensure that the scientific evidence produced by the experts appointed to the Independent Verification Group (IVG) was accepted by the negotiating parties. Disagreements about areas of evidence occurred less frequently after the IVG assessment. This suggests that gaining explicit agreement on the assessors and areas to be assessed, while having limitations (Schirmer, Dare & Ercan, 2016), assisted in enabling the evidence to be used constructively in subsequent negotiations, rather than becoming another area of disagreement among participants.

Principle 3: Agree How to Deal with Unknowns

Scientific understandings of environmental systems are constantly evolving. Scientific evidence and knowledge that is current at a given point in time will almost inevitably be superseded as land and water use changes, and as scientific knowledge evolves. Attempts to resolve environmental conflicts at a given point in time are unlikely to succeed unless a mechanism is included for monitoring and evaluating outcomes, and for reviewing what has been agreed to as new evidence emerges; this means including a process for how to address 'unknowns'—that is, factors that will likely emerge in the future—as part of any agreement reached about an environmental conflict.

The RFAs dealt with this by requiring forest management systems to be 'capable of responding to new information' (Commonwealth of Australia and State of Tasmania, 1997), and by having a five-yearly review that evaluated whether the systems were meeting the requirements of the RFA (including responding to new information). The Murray–Darling Basin Plan has similar requirements for monitoring and evaluation. However, in both these processes, there is limited ongoing funding for evaluating emerging scientific evidence, which is likely to limit the effectiveness of these review and monitoring processes in addressing new and emerging issues. This is a key issue, for when new concerns and evidence emerge, fresh conflict will almost inevitably arise unless there is a transparent process available by which these can be evaluated or addressed—in other words, a space in which stakeholders can raise and discuss issues that were unknown when the original agreement was reached.

Recommendations

Successfully managing land use in Australia requires robust systems for addressing environmental conflict in constructive and appropriate ways. A critical part of this is ensuring that scientific evidence contributes to achieving greater consensus, rather than further deepening and dividing opinions about how best to manage Australia's diverse environments. The following three principles can assist in doing this:

- Agree on your values: How will the science be interpreted? What values will be used to determine thresholds of acceptable social, economic and environmental impacts of land use change? Agreement on shared values reduces the potential for scientific evidence to be interpreted differently by different groups.
- Agree on how you will do the science: Explicitly agree on who will evaluate scientific evidence. Which scientific experts are accepted by all those involved in an environmental conflict? What criteria will be used to judge the quality, validity and reliability of scientific evidence?
- Agree on how to deal with unknowns: How will emerging and new evidence be identified, examined and responded to into the future?

Successfully implementing these principles requires constructive dialogue between all parties involved in the conflict. They will not assist in the interpretation of scientific evidence if some parties' views have not been included in the discussion of, and agreement on, the principles. Their success rests on ensuring that those involved in a conflict have the time and resources to discuss issues and agree on principles. Successfully bringing scientific evidence to bear in environmental conflicts requires constructive and inclusive dialogue between all parties.

References

Brueckner, M. & Horowitz, P. (2005). The use of science in environmental policy: A case study of the regional forest agreement process in Western Australia. *Sustainability: Science, Practice, & Policy 1*(2), 14–24.

Commonwealth of Australia and State of Tasmania. (1997). *Tasmanian regional forest agreement*. Canberra, ACT: Commonwealth of Australia.

Daniell, K. A. (2011). Enhancing collaborative management in the basin. In D. Connell & Q. Grafton (Eds.), *Basin Futures: Water reform in the Murray-Darling Basin* (pp. 413–37). Canberra, ACT: ANU E Press.

Furlong, G. T. (2010). *The conflict resolution toolbox: Models and maps for analyzing, diagnosing, and resolving conflict.* Hoboken, NJ: John Wiley & Sons.

Grafton, Q. R. & Jiang, Q. (2011). Economic effects of water recovery on irrigated agriculture in the Murray–Darling Basin, *Australian Journal of Agricultural and Resource Economics* 55(4), 487–99. doi.org/10.1111/j.1467-8489.2011.00545.x

Harding, R., Hendriks, C. M. & Faruqi, M. (2009). *Environmental decision-making: Exploring complexity and context.* Sydney, NSW: Federation Press.

JANIS (Joint ANZECC/MCFFA National Forest Policy Statement Implementation Sub-committee). (1997). *Nationally agreed criteria for the establishment of a comprehensive, adequate and representative reserve system for forests in Australia.* Canberra, ACT: Australian Government Publishing Service.

Kirkpatrick, J. B. (1998). Nature conservation and the Regional Forest Agreement process. *Australian Journal of Environmental Management* 5(1), 31–7. doi.org/10.1080/14486563.1998.10648397

Longino, H. E. (1990). *Science as social knowledge: Values and objectivity in scientific inquiry.* Princeton, NJ: Princeton University Press.

Mercer, D. (2000). *A question of balance: Natural resources conflict issues in Australia.* Sydney, NSW: Federation Press.

MDBA (Murray–Darling Basin Authority). (2014). *The journey to a basin plan—an overview.* Canberra, ACT: Murray–Darling Basin Authority.

Musselwhite, G. & Herath, G. (2005). Australia's regional forest agreement process: Analysis of the potential and problems. *Forest policy and Economics* 7(4), 579–88. doi.org/10.1016/j.forpol.2003.11.001

Poynter, M. (2013). Response to Martin Moroni's guest editorial in Australian forestry. *Australian Forestry* 76(1), 69–70. doi.org/10.1080/00049158.2013.776929

Quiggin, J. (2012). Why the guide to the proposed basin plan failed, and what can be done to fix it. In J. Quiggin, S. Chambers & T. Mallawaarachchi (Eds.), *Water policy reform lessons in sustainability from the Murray–Darling Basin* (pp. 49–60). Cheltenham, UK: Edward Elgar. doi.org/10.4337/9781781000328.00014

Regional Development Australia Northern Inland. (n.d.). *Measuring the socio-economic impacts of the MDB Plan.* Retrieved from www.rdani. org.au/our-region/current-regional-issues/socioeconomic-impacts-of-the-murray-darling-basin-plan.php

Schirmer, J. (2013). Engaging with scientific data: Making it meaningful. In H. J. Aslin & S. Lockie (Eds.), *Engaged environmental citizenship* (pp. 87–105). Darwin, NT: Charles Darwin University Press.

Schirmer, J., Dare, L. & Ercan, S. (2016). Deliberative democracy and the Tasmanian forest peace process. *Australian Journal of Political Science 51*(2), 288–307. doi.org/10.1080/10361146.2015.1123673

West, J. (2012). *Final report on the work of the independent verification group for the Tasmanian forests Intergovernmental agreement.* Retrieved from www.environment.gov.au/resource/independent-verification-group-report

Wynne, B. (2006). Public engagement as a means of restoring public trust in science—hitting the notes, but missing the music? *Community Genetics 9*, 211–20. doi.org/10.1159/000092659

Young, W. J., Bond, N., Brookes, J., Gawne, B. & Jones, G. J. (2011). *Science review of the estimation of an environmentally sustainable level of take for the Murray–Darling Basin.* Retrieved from www.mdba. gov.au/sites/default/files/archived/proposed/CSIRO_ESLT_Science_Review.pdf

12

The Role and Importance of Coordinated Land Information to Support Landscape Connectivity Initiatives

Gary Howling and Ian Pulsford

Key Points

- 'Connectivity conservation' engages participants from a range of sectors of society and encourages the collaboration and alignment of efforts to conserve and connect habitats across whole landscapes.

- The Great Eastern Ranges Initiative (GER) was established as a connectivity conservation program to facilitate collaborative, cross-tenure conservation management along the 3,600 kilometres of mostly interconnected natural lands that extend along the Great Dividing Range and Great Escarpment of eastern Australia.

- The scale of the GER vision and the need to achieve a range of ecological, social and institutional outcomes across such a large area created new challenges for managers seeking to understand and plan for large-scale processes.

- At the continental scale, GER managers needed to understand the landscape context before being able to make clear decisions about where, how and when to act. This required the collation and analysis of information on factors such as the distribution, connectedness and adequacy of existing protected areas, their role in providing seasonal drought and long-term climate refuge, their use by migratory species and predicted changes in habitat condition into the future.

- Combining these analyses allowed the GER to develop a strategic overview of how each region within the broader landscape contributes to the status of the whole.

Introduction to 'Connectivity Conservation'

The accelerating loss of biodiversity from a range of threats, including climate change and land clearing, has stimulated a social and political shift in the management of global biodiversity (Worboys & Pulsford, 2011). Experience in various countries has shown that conservation managers need to move beyond the traditional approach of conserving isolated pockets of habitat to ensure the long-term survival of species and ecosystems (see Chester, 2006; McKinney, Scarlett & Kemmis, 2010; Soulé & Terborgh, 1999). Australia has a long history of investing in public and private efforts to manage natural areas that help to maintain natural and cultural heritage in the face of growing pressures. This has been the case in the highly managed landscapes of the eastern states (including Tasmania) and south-west Western Australia. Attempts to expand, link and buffer a network of protected habitats in these regions have been increasingly targeted and coordinated through enterprises variously referred to as 'wildlife corridors', 'biolinks' or 'integrated catchment management areas' (Department of Environment and Climate Change [DECC], 2008; Department of Sustainability, Environment, Water, Populations and Communities, 2012; Mackey, Ferrier & Possingham, 2013).

Connectivity conservation is a recent response to the need for a more expansive whole-of-landscape effort. It involves the active conservation management of natural and semi-natural areas across a range of land tenures and uses (Worboys, 2010). The purpose of connectivity conservation is to:

- conserve natural vegetation and habitats
- interconnect protected areas and other natural lands and permit the movement of animals and plants between them
- conserve animal species and healthy ecosystem processes

- respond to climate change by providing opportunities for species to move to locations with suitable climate envelopes (including altitudinally) as local environmental conditions change (International Union for the Conservation of Nature [IUCN] – World Commission on Protected Areas, 2006).

Unlike traditional single-landscape or multi-partner projects, connectivity conservation is undertaken by individuals, communities, private organisations and governments working together as a cohesive enterprise, seeking to achieve goals and objectives agreed to voluntarily by participants. The work is typically undertaken on a voluntary basis and is guided by a clear overall vision for a large-scale corridor (Worboys et al., 2016). By necessity, it involves making provision for natural processes at all spatial scales to accommodate the needs of local species, facilitate migration, maintain healthy ecosystem processes and increase the resilience of habitats. Connectivity conservation involves:

1. conservation management on lands around formal protected areas to buffer them from threatening processes originating off-reserve

2. large-scale ecological restoration and rehabilitation on heavily cleared lands to reconnect otherwise isolated protected areas

3. management and suppression of processes that would otherwise degrade the values of largely intact, high-conservation value habitat and wilderness

4. systematic conservation planning to factor in the management needs of large-scale, spatially dependent ecological processes essential for the long-term persistence of biodiversity (Mackey, Watson & Worboys, 2010).

The remainder of this paper examines the last of these considerations and focuses on how the analysis of large-scale ecological processes were considered in strategic decisions made as part of the Great Eastern Ranges Initiative (GER).

The Great Eastern Ranges

Australia's iconic Great Eastern Ranges span 3,600 kilometres (2,237 miles) from the Grampian Ranges in western Victoria, along the Great Dividing Range, through Queensland's World Heritage wet tropics to the remote peninsula of Cape York. In total, the landscape occupies some 33 million hectares (GER, 2015). From undulating heath-covered slopes to the towering slopes of Mount Kosciuszko, the Great Eastern Ranges are a biodiversity hotspot, rich in natural resources and cultural associations that are highly valued by Aboriginal and non-Aboriginal Australians. The cultural, biodiversity and ecosystem service values of the Great Eastern Ranges have been well documented. They include:

• the longest range of mountainous and upland landscapes on the continent, spanning 21 degrees of latitude and including our greatest altitudinal gradient

• the most reliable source of water, providing fresh water for at least 11 million people across eastern Australia, both on the coast and across the inland catchments

• the greatest variety of habitats and species, including globally significant hotspots for species diversity and endemism, and habitat for 60 per cent of Australia's threatened animals and 70 per cent of its plants

• ancient species like the Wollemi Pine and ancient flowering rainforest plants, which provide living connections to our deep geological history

• migration pathways supporting the annual seasonal dispersal and long-distance movement of up to 60 per cent of Australia's forest- and woodland-dependent birds, such as the rainbow bee-eater (*Merops ornatus*) and regent honeyeater (*Xanthomyza phrygia*), as well as iconic Australian species such as the grey-headed flying fox (*Pteropus poliocephalus*), bogong moth (*Agrotis infusa*) and Richmond birdwing butterfly (*Ornithoptera richmondia*)

• an extensive network of more than 2,000 existing protected areas on public and private lands, which provides the basis for seeking to achieve the GER vision (Australian Conservation Foundation [ACF], 2015; Dean-Jones, 2009; Hyder Consulting, 2008; Williams et al., 2011).

A complex mix of ongoing pressures similarly affects the Great Eastern Ranges, exacerbating the effects of past and new disturbances that cause habitat loss. These have been documented extensively (see Great Eastern Ranges, n.d.) and include widespread agricultural use on the more fertile soils, logging, grazing, mining, urban development, competition from introduced species and changed disturbance regimes (e.g. fire and hydrology). Across extensive parts of the Great Eastern Ranges, various pressures have combined to exacerbate the loss or fragmentation of habitat for an alarming proportion of plant and animal species that were formerly more widespread (ACF, 2015; Mackey et al., 2010). Despite this, the Great Eastern Ranges provide Australia's best opportunity to resist both the combined pressures of a growing human population encroaching from the east (DECC, 2007, 2008; Mackey et al., 2010) and the effects of climate change, which are forcing environmental envelopes to the south-east in geography and upwards in elevation (Doerr et al., 2013).

The Great Eastern Ranges Initiative

The New South Wales (NSW) Government established the GER in 2007 as a long-term strategy to enhance the health and connectivity of a network of natural ecosystems across eastern Australia, and mitigate increasing threats to the values they contain. Specifically, the GER was established to implement connectivity conservation across multiple state and territory jurisdictions as part of a vision to 'bring people and organisations together to protect, link and restore healthy habitats over 3,600 kilometres from western Victoria through NSW and the ACT to far North Queensland' (GER, 2012) (see Figure 12.1).

Connectivity conservation planning and on-ground conservation activities are structured to achieve four goals (see Table 12.1). The first relates to the 'on-ground' outcome that will be achieved in relation to connecting landscapes and ecosystems. Complementary goals direct delivery through a partnership approach, communicating with the wider community to increase awareness and support active participation and the application of knowledge. Delivery is achieved by building on a foundation of collaborative public–private partnerships established since 2007. Ten regional partnerships provide a focus for local action, bringing together landholders, agencies, non-government organisations, community and Indigenous groups, researchers, councils and industry to collectively plan and carry out projects. In addition, a number of well-recognised regional, state and national organisations have chosen to align their own activities with the vision of the GER. By leveraging the combined power and knowledge of partner organisations and regional partnerships, the GER has expanded its presence to create corridors of effort encompassing the full extent of the Ranges.

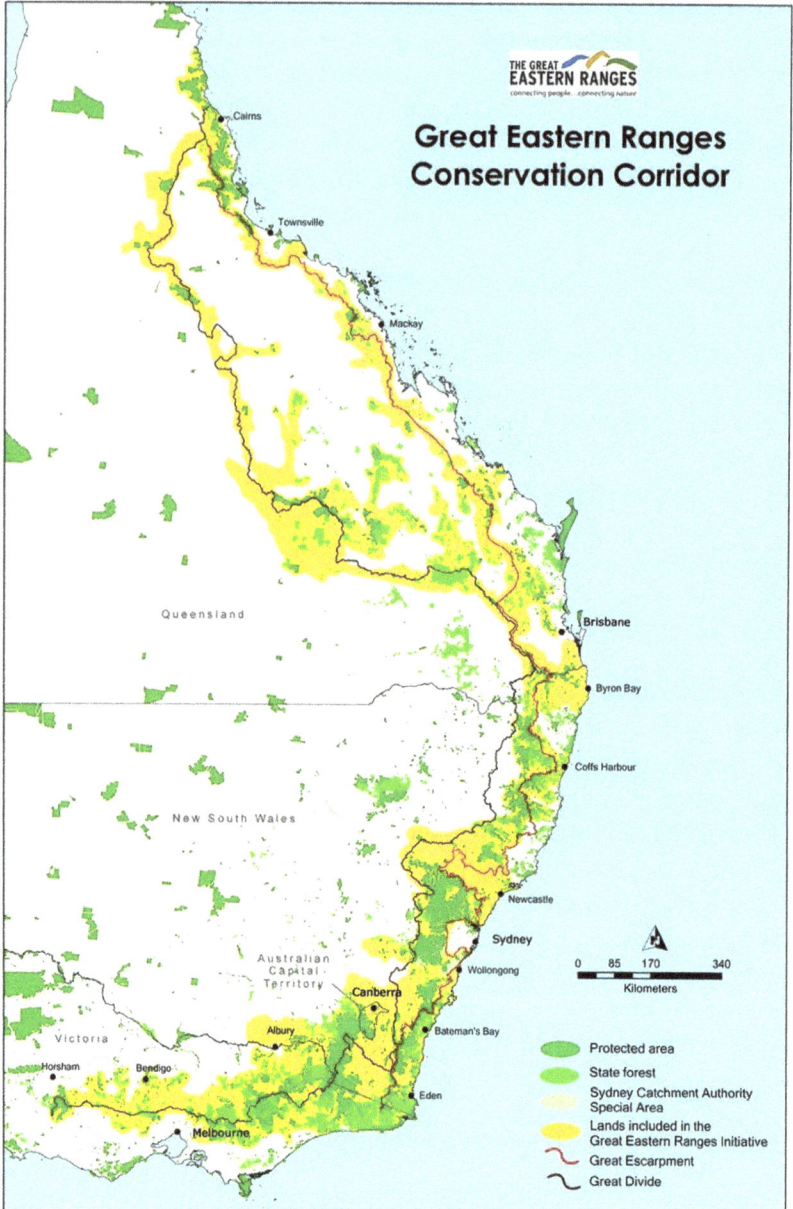

Figure 12.1: Map of Australia's Great Eastern Ranges.
Source: Pulsford, Worboys & Howling (2010).

Table 12.1 Goals of the Great Eastern Ranges Initiative

Theme	Goal
Connectivity conservation	Spanning from Cape York in Queensland to Walhalla and the Grampians in Victoria, the Great Eastern Ranges corridor is maintained as a biodiverse, functionally interconnected connectivity conservation area that positively contributes to biodiversity conservation and the delivery of ecosystem services.
Leadership and governance	Leadership and governance stewardship by the GER Board deliver effective and financially sustainable continental-scale connectivity conservation actions for the GER corridor.
Management	The GER corridor is actively and effectively managed: threats are responded to and restoration contributes to connectivity conservation, healthy ecosystems, biodiversity conservation and landscape amenity. The GER corridor provides direct service to the Australian community as a whole-of-continent natural response to climate change with active management facilitating adaptation and mitigation benefits.
Community awareness, involvement and partnerships	The GER corridor is a highly regarded household name throughout the connectivity conservation area, and the corridor is positively supported by communities. The GER corridor vision is actively managed by multiple positive, sustained and cooperative contributions from individuals, communities, private organisations, government organisations and other stakeholders.
Information	GER managers and the GER community are consistently informed by the very latest monitoring, research and modelling information that provides conditions and trends in condition analysis, adaptation research results, forecasting information, evaluation of performance results and other critical GER corridor management information.

Source: GER (2016).

The Role of Regional Partnerships

From the earliest days of the GER, it was understood that a whole-of-landscape approach to conservation would be essential to conserve the landscape's unique species, facilitate their migration, maintain ecosystem processes and increase the resilience of habitats (DECC, 2007, 2008). The GER's scope and vision inspired many organisations, landowners and managers to contribute to planning and implementing on-ground conservation actions. This was done in a coordinated and targeted way that has helped to increase public support for biodiversity conservation activities.

A collaborative approach was used to establish 'regional partnerships' that brought together landholders, public land managers, non-government conservation organisations, councils and other stakeholders to coordinate their efforts towards commonly agreed goals and targets (GER, 2012). The GER regional partnership delivery model has encouraged participating organisations in each priority region to agree on common values and objectives as the basis for collaborative local action, and to develop plans to prioritise actions in locations that contribute most strongly to connectivity at local and regional scales. Regional partnerships were also vital in providing a framework to develop locally appropriate governance structures and frameworks to enable the involvement of a diverse range of contributors and supporters from all sectors.

On a practical level, the partnership model enabled the GER to mobilise a broader cross-section of public and private land management approaches across all tenures (GER, 2012), including:

- protection of habitat on private land through a spectrum of conservation options ranging from entry-level instruments to in-perpetuity conservation covenants
- collaborative management of invasive weeds and feral animal populations
- habitat restoration, with emphasis on mobilising a range of techniques to enhance the condition of remnant areas and the functional connectivity value of intervening areas
- community engagement, landholder capacity building and promotion of ways to get involved
- provision of access to a wide range of information and reports—for example, through the Great Eastern Ranges website.

Prioritising Investment and Action

Considerable time, effort and investment are required to unite a group of partners, agree on common priorities and plan a program of works. Around 80 per cent of the GER corridor is comprised of lands other than protected areas (Hyder Consulting, 2008), indicating a range of opportunities in which efforts could be targeted. A clear rationale for identifying areas requiring connectivity investment was essential for confirming the scope and direction for activities to partners in priority regions, and demonstrating transparency in program-level investment decisions to people in other regions (GER, 2012).

In mid-2009, the first GER Science and Information Delivery Plan 2010–2015 was developed by an independent panel. The plan highlighted the vital need for science-based information to depict landscape condition and guide planning at a whole-of-corridor scale. Many spatial analyses were completed including:

- a map of the distribution of existing public protected areas and private land conservation instruments
- an analysis of landscape-scale connectivity of habitats of eastern Australia (Drielsma, Barrett, Mannion & Love, 2010)
- an assessment of the distribution and significance of potential drought and climate refuge areas (Mackey & Hugh, 2010)
- an assessment of seasonal bird migration and dispersal routes (Smith, 2010)
- an assessment of the current condition of remnant vegetation and potential for future loss of condition at the landscape scale throughout the GER corridor (Drielsma, Howling & Love, 2010).

Several of these spatial layers were used to describe ecological processes and patterns operating at large regional scales, and to identify areas for targeting on-ground conservation investment and management on a local scale. A high-level overview of how these informed understandings of the status and priorities for investment at the GER scale is provided below.

The management and analysis of large-scale processes proved challenging for the GER from the outset, as it required new ways of thinking and approaches to understanding and planning for processes and activities operating across extensive areas. The next section outlines how the various analyses were used individually and collectively to inform the selection of priority landscapes to be targeted for the formation of new regional partnerships. The processes used in the analyses are not described; these are well documented in the relevant report findings and in Howling (2013).

The Distribution of Existing Public Protected Areas and Private Land Conservation Instruments

Protected areas on public and private land form both the foundation and core for any connectivity conservation initiative (Soulé & Terborgh, 1999; Worboys et al., 2016). Previous assessments of conservation values in protected area networks by state conservation agencies in Queensland, NSW, Victoria and the Australian Capital Territory (ACT) highlighted limitations in the existing reserve system (e.g. Government of Australia, 2007; National Parks and Wildlife Service, 2003; Sattler & Williams, 1999). However, the GER needed a more complete understanding. In 2010, it commissioned a spatial analysis of all public land tenures (arranged as IUCN protected area management categories) and private land conservation instruments (conservation covenants and private reserves) that form the core connectivity conservation network of the GER corridor.

Initially, the map and associated database were intended to provide a baseline against which it might be possible to track changes in patterns of reserve establishment and uptake in private land conservation instruments. However, it also proved important in providing a snapshot of the status of connectivity conservation at that time. Several patterns emerged that highlighted the general extent and connectivity of the protected area network. This was used to identify gaps in connectivity at regional, state and continental scales in different parts of the GER corridor. The analysis revealed some very substantial core sections of contiguous protected areas, including 'large intact landscapes' protected in contiguous national parks and other reserves, such as those extending from around the Australian Alps and Greater Blue Mountains to the Border Ranges. These act as core areas at the heart of the GER corridor in Victoria, ACT and NSW, but not in Queensland, which has a network of substantially fewer and less-extensive interconnected protected areas. Outside these core areas are 'tenure mosaic landscapes', which comprise a mix of public protected areas and other public lands, interspersed with private lands. Native habitats in these landscapes generally remain fairly intact, such as those across extensive areas of northern Queensland and northern NSW. However, the security and long-term viability of connections within these areas are not assured and they may be liable to degrade in future. Areas identified as 'conservation gaps', which comprise either very narrow connections or gaps in the connectedness of conservation instruments, are generally adjacent to landscapes that have been extensively cleared and developed in the past. Many of these are subject to a wide range of threats.

Implementing a strategic integrated approach to conservation management in each of these connectivity management contexts was recognised as important. This can be achieved by collaborative cross-tenure management of invasive species along the margins of protected areas and targeted delivery of long-term and in-perpetuity conservation agreements to link public protected areas (Howling, 2012).

Landscape-Scale Connectivity of Habitats

Analysis of the pattern of conservation instruments was complemented by analysis of the connectivity of forest and woodland habitats through and between extant woody vegetation. The modelling (Drielsma, Howling & Love, 2010) was undertaken at a range of scales from continental to priority area (e.g. Upper Hunter Valley and slopes to summit).

In addition to assisting planning decisions, the production of maps depicting connectivity of habitats proved to be a valuable tool: first in communicating the status of existing linkages, then in demonstrating the significance of the local landscape relative to the wider GER corridor. This motivated landholders, community groups, Landcare and others to feel part of 'the bigger picture', which, in turn, led to higher rates of involvement by landholders in private land conservation agreements and habitat restoration, as well as greater alignment of delivery organisations' priorities and targeting of resources (Dunn & Howling, 2015).

Drielsma, Barrett et al. (2010) highlighted numerous significant patterns that needed to be considered by the GER in planning the placement of future regional partnerships, including:

- the role of large intact reserved landscapes in contributing to connectivity of the GER corridor on a continental scale
- landscape-scale connections between the GER corridor and adjacent landscapes, including the 'western woodlands way' network of dry forests and woodlands on the inland slopes and plains, and lowland habitats along the NSW coast; these linkages range from local-scale connections within and along the fringe of the GER corridor, to major connecting landscapes

- Hanging Rock to Watsons Creek and NSW North West Slopes
- Coffs Coast to Dorrigo and Barrington–Comboyne to Queens Lake (north and central coasts)
- Liverpool Range, Winburndale, Kanangra Boyd–Wyangala (NSW Central Tablelands) and NSW South West Slopes

- 'bottlenecks' whereby connectivity is narrowed because of a natural constriction of the ranges, often exacerbated by surrounding land use; these landscapes (such as Malanganee to Richmond Ranges, Illawarra Escarpment and upper Bega Valley) represent a significant potential risk for lost connectivity on a continental scale, and in some cases (e.g. Upper Hunter Valley and NSW Southern Highlands) have provided a focus for effort in the GER

- connectivity gaps, where land use development physically bisects the latitudinal or altitudinal connectivity of the GER corridor; these include settlement and infrastructure corridors associated with major crossings, such as the Bruxner, Golden, Great Western and Hume highways, and developed areas, such as the Malanganee Gap, Blue Mountains, NSW Southern Tablelands and ACT landscapes.

In summary, the analysis of connectivity of the GER habitats was, in itself, highly informative in pointing to significant existing or potential weaknesses in connectivity of the GER corridor as a whole, particularly where these occurred in landscapes with few public or private protected areas.

Seasonal Bird Migration and Dispersal Routes

Maintenance of local- and landscape-scale connections will be essential to ensure their continued contribution to seasonal migrations, and species dispersal and adaptation following major climatic changes or landscape disturbance events (e.g. fires, droughts and floods). One of the GER's earliest commissioned science projects involved exploring the migration pathways used by birds. Birds are just one group of species that use the GER corridor; they are not necessarily fully representative of all types of movement seen in the corridor. However, they are an extremely useful focus for understanding habitat use and connectivity because:

- they are mobile
- they already provide the focus for extensive community action

- the Birdlife Australia 'Birdata Atlas' and network of community observers provide an extensive dataset to draw from
- in Birdlife Australia, the GER already had a partner organisation with the knowledge and ability to interpret what the analyses were observing.

In approaching the dataset, the GER specifically wanted to know:

- Where do they occur throughout the course of the year?
- How do they use the GER corridor relative to surrounding landscapes?
- Are there particular pathways that are more important for movement?
- Are there any noticeable barriers to movement at this large scale?

The existence of functional links between habitats are of particular importance for the movement of migratory birds, the dispersal of fledgling birds and birds responding to changes in their environment, such as climate change or fires. Analysing observational records in the Birdata Atlas for 18 species with recognisably seasonal migration patterns, or long-distance dispersal, enabled the movement of birds to be tracked along clearly defined pathways—referred to as the 'flyways' of eastern Australia (Smith, 2010).

Five general patterns of movements were revealed within the GER corridor, reflecting variation in how different species use the landscape as a result of their habitat preferences and gap-crossing ability:

1. Broadscale latitudinal migrants (e.g. dollarbird [*Eurystomus orientalis*] and rainbow bee-eater [*Merops ornatus*]): these species are considered to have more general habitat requirements, allowing them to migrate across a broad sweep of the GER corridor using both forested and cleared areas, and a variety of forest and woodland types.
2. Restricted-scale latitudinal migrants (e.g. rufous fantail [*Rhipidura rufifrons*] and satin flycatcher [*Myiagra cyanoleuca*]): these species are restricted to vegetated areas along the GER corridor and coastal regions. They appear to be the most sensitive to the effects of fragmentation and other forms of habitat degradation and show where bottlenecks and habitat gaps exist within the GER corridor.

3. Partial migrants (e.g. grey fantail [*Rhipidura albiscapa*] and dusky woodswallow [*Artamus cyanopterus*]): these are the species for which short-distance migrations are most commonly noted; however, for the time-series analysis, they were not useful in showing such effects.

4. Altitudinal migrants (e.g. flame robin [*Petroica phoenicea*] and gang-gang cockatoo [*Callocephalon fimbriatum*]): these species undertake short migrations characterised by a shift between higher elevations within the GER corridor and lower elevations of the adjoining coastal lowlands or inland slopes.

5. 'Rich-patch' migrants (e.g. swift parrot [*Lathamus discolour*] and regent honeyeater [*Anthochaera Phrygia*]): these species display some regularity in their movement in time-series data; however, their routes were less evident than those of other species, as they are particularly rare and many of their movements depend on the distribution of flowering trees.

The most notable bird movements were the range contractions into Queensland and between the high country and inland coastal plains for overwintering. These patterns of movement highlight areas of high and low connectivity within the GER corridor at local and continental scales. For example, migration routes used by rufous fantail appeared to be bottlenecked in the regions east of the Hunter Valley and NSW Southern Highlands. The protection and restoration of habitat corridors that provide functional connectivity across the landscape of the GER corridor are essential for the long-term viability of many species.

Several landscapes were highlighted as potentially forming bottlenecks in seasonal bird migration—specifically the:

- western arc of the Border Ranges and the coastal lowlands flyway through the Gold Coast hinterland leading north into Queensland
- Mallanganee and Richmond Ranges and Big Scrub linkages leading south from the NSW–Queensland border
- Upper Hunter Valley (centred on four riparian corridors across the Merriwa Plateau and Manobalai Range) and Lower Hunter coastal corridor
- NSW Southern Highlands and Illawarra Escarpment
- northern ACT–NSW Southern Tablelands, linking Alpine reserves with the Mundoonen Range and NSW Southern Tablelands flyway (in turn, part of a western flyway into northern Victoria)

- upper Bega Valley (comprising part of the coastal flyway linking with northern Victoria).

As with the depiction of habitat connectivity, demonstrating the visual correlation between where linkages exist in the landscape and where birds are moving on a seasonal basis was highly useful in planning future priorities and engaging communities and delivery partners in understanding the relative importance of each part of the GER corridor for bird migration.

Current and Potential Future Vegetation Condition

Vegetation condition is often used as a surrogate for the proportion of biodiversity that a site can potentially support. It provides an indication of 'effective habitat area'—that is, the area that can support targeted species of the maximum set of biota normally associated with an ecosystem (Drielsma, Barrett et al., 2010). It is also an important factor influencing functional connectivity values—that is, supporting effective species movement and ecological interactions and some ecosystems (e.g. carbon capture, water quality and catchment yield).

Across the GER corridor and adjacent landscapes, a baseline measure of vegetation condition was defined at 'supra-regional' scale, based on the interaction of land tenure (influences, management and security), land cover (contrasting cleared areas with extant native vegetation), land use (accommodating differences in the type and intensity of pressures expected from different uses) and changes in vegetation structure (based on changes in canopy density relative to 'benchmarks' for each woody vegetation type).

Across the GER corridor, the modelling highlighted at least 10 significant gaps in functional connectivity at the whole-of-GER scale resulting from clearing and other factors. In addition, it was noted that gaps formed barriers to movement between the GER corridor and natural ecosystems in adjacent landscapes, such as the NSW Southern Tablelands and peri-urban landscapes adjoining major cities. This affected the potential for movement into and out of the GER corridor for species other than habitat generalists.

Scenarios for future condition were modelled to accommodate the likely effects of key threats associated with increased land use pressure (affecting permeability of the landscape matrix), increased human population density (implying increased habitat fragmentation and likelihood of disturbance) and proximity to infrastructure (as a surrogate for past disturbance, and likelihood of invasion by exotic plants, feral fauna and ignition of wildfires).

Combining Results to Inform Connectivity Conservation Priorities

Each of the analyses described above proved informative by themselves. However, their value for demonstrating the potential for landscape patterns, ecological processes and human pressures became more apparent when viewed together (Dunn & Howling, 2015). The GER trialled an approach to prioritising each section of the GER corridor in NSW to guide decisions regarding future efforts (Howling, 2012). The spatial products discussed above were applied to derive an understanding of each landscape within the GER corridor. Subsequent analysis identified a series of focus landscapes based on four criteria:

1. biological values—considering the contribution made at regional and continental scales to the 'fabric' of the GER corridor and the ecological processes it supports, including regional distinctiveness and species diversity, in situ resilience of ecosystems and native species, ecosystem processes and climate adaptation potential

2. connectivity need—considering the apparent discontinuity in connectedness of habitat and protected areas, potential for current functional connectivity to be diminished or lost, or potential for current gaps in functional connectivity to be made worse or less retrievable

3. social and institutional capacity—considering opportunities presented to implement a connectivity conservation initiative, based on there being active organisations present in the landscape with the capacity and interest in contributing to delivery of collaborative projects

4. program contribution—exploring opportunities presented to develop and test approaches that contribute to implementing an effective GER, and delivering outcomes in relation to planned geographic expansion within the GER corridor.

The analysis highlighted the extent to which each section of the GER corridor contributes to the corridor as a whole and relative to each other. In doing so, it highlighted the existence of a number of gaps or weaknesses in connectivity of the GER corridor. These included core habitat areas under sustained pressure from edge effect, habitat areas that form natural and fragmentation-derived bottlenecks, species dispersal or migration, and landscapes with potential for continued loss associated with the erosion of functional connectivity.

Based on the results of this process, the 2011–15 period saw the formation of five new partnerships in the GER corridor, including three in NSW and one each in Queensland and Victoria:

- Sunshine Coast Hinterland Bushlinks (established January 2012): this partnership area is located within a recognised biodiversity hotspot centred on the Glasshouse Mountains; it supports an important linkage between the Blackall and Conondale Ranges where weed management has been the focus of the group's activities.

- Jaliigirr Biodiversity Alliance (established May 2012): formed as an incorporated entity, this partnership area covers 337,000 hectares from Coffs Coast to the Dorrigo Plateau and is located in a tropical, subtropical and temperate convergence. The region is an area of significant ecological diversity and includes the World Heritage Gondwana Rainforests of Australia.

- Kanangra Boyd to Wyangala Link (established August 2012): this partnership was formed to implement a major $2.7 million landscape connectivity project on NSW's Central Tablelands; it was funded by the Australian Government's Clean Energy Futures Biodiversity Fund until June 2017.

- Illawarra to Shoalhaven (established October 2012): this partnership is located where the Illawarra and Cambewarra Escarpments combine to form a narrow north–south aligned rainforest corridor linking the major sandstone reserves of the southern Sydney Basin with the wet sclerophyll forests of the NSW south coast.

- Central Victorian BioLinks (established 2013): this partnership is located between the Grampians and West Gippsland in central Victoria; it spans an ecological gradient from drier northern plains, across the woodlands and forests of the divide, to cooler and more southerly hills, gorges and grasslands.

The analyses carried out in 2010, combined with lessons learned in relation to social considerations in partnership formation (stemming from five original regional GER partnerships), proved essential in assisting the successful establishment of each of these new partnerships. The existing partner organisations and the five new regional partnerships are committed to acting cooperatively in each of these landscapes. Together, they seek:

- increased knowledge and understanding of the biological attributes of the Great Eastern Ranges, and their significance in the context of broader continental ecosystems
- greater awareness and understanding of biological, human and existence values of the GER corridor, threats to these values and opportunities and priorities for action to address these
- increased recognition of the scientific credibility and validity of connectivity conservation in the Eastern Ranges, and the importance of continued investment to understand and address conservation priorities
- increased adoption of advice from scientists and research by public and private land managers, and the mainstreaming of GER-focused conservation priorities in partners' strategic planning and program implementation.

The Role of Coordinated Land Information into the Future

In 2015, a new science and information plan (GER, 2015) was prepared to facilitate the acquisition and use of appropriate data, information and knowledge by the GER partners. This information can be used to mainstream GER-focused conservation priorities in strategic planning and investment. To better understand management needs and priorities, the plan identified five questions for consideration:

1. How is connectivity conservation important for biodiversity conservation and supporting ecosystem services in the GER corridor?
2. What are the priority regions in the GER corridor where we should focus connectivity conservation management?
3. Within priority regions, what conservation outcomes should we seek to achieve, where should we act and how should we act to be most effective?

4. How should we bring people and institutions together to deliver ethical and effective GER governance, planning and management?

5. Are we being effective in achieving desired outcomes across all relevant spatial scales?

The conservation challenges and information requirements to achieve the GER's ambitious vision are complex and manifold. Data and information at the scale of the GER corridor will be increasingly vital to addressing these questions and ensuring the successful delivery of collaborative efforts. Access to sound information on the management needs of native species and ecosystems, and the processes and pressures acting upon them, will remain essential in supporting prioritisation monitoring and evaluation of threats, investment in conservation actions and delivery outcomes (Mackey et al., 2010; Office of Environment and Heritage, 2011).

Since 2007, the GER has evolved and grown to become one of the largest connectivity conservation initiatives of its kind in the world, with a diverse cross-section of participants operating across extensive areas of the corridor. This success has depended substantially on the quality of scientific information that has been used to stimulate and inform planning, set priorities and build community engagement and support. The experience of the GER demonstrates that having access to (and a strategy to use) well-coordinated information at a range of scales is essential if connectivity conservation initiatives are to realise a landscape vision for large-scale connectivity.

References

ACF (Australian Conservation Foundation). (2015). *Our great dividing range: Restoring life to our heartland* [Australian Conservation Foundation website]. Retrieved from stories.acf.org.au/our-great-dividing-range

Chester, C. C. (2006). *Conservation across borders: Biodiversity in and interdependent world.* Washington, DC: Island Press.

Dean-Jones, P. (2009). *High country heritage—cultural heritage themes for the GER Initiative in NSW.* Sydney, NSW: Umwelt (Australia) Pty Ltd.

DECC (Department of Environment and Climate Change). (2007). *The Alps to Atherton initiative: NSW business plan.* Queanbeyan, NSW: Department of Environment and Climate Change.

DECC. (2008). *A new biodiversity strategy for New South Wales: Discussion paper.* Sydney, NSW: Department of Environment and Climate Change.

Department of Sustainability, Environment, Water, Populations and Communities. (2012). *National wildlife corridors plan: A national framework for landscape-scale conservation.* Canberra, ACT: Department of Sustainability, Environment, Water, Populations and Communities. Retrieved from www.environment.gov.au/biodiversity/wildlife-corridors/publications/national-plan.html

Doerr, V. A. J., Williams, K. J., Drielsma, M., Doerr, E. D., Davies, M. J., Love, J., ... Ferrier, S. (2013). *Designing landscapes for biodiversity under climate change: Final report.* Gold Coast, QLD: National Climate Change Adaptation Research Facility.

Drielsma, M., Barrett, T., Mannion, G. & Love, J. (2010). *Spatial analysis of conservation priorities in the great eastern ranges, project 1, vegetation condition.* Armidale, NSW: Department of Environment, Climate Change and Water.

Drielsma, M., Howling, G. & Love, J. (2010). *Evaluating native vegetation management benefits in New South Wales: Technical report.* Armidale, NSW: Office of Environment and Heritage, Department of Premier and Cabinet.

Dunn, R. & Howling, G. M. (2015). *The Great Eastern Ranges initiative 2011–2015: Building for the future.* Sydney, NSW: Greening Australia and Office of Environment and Heritage.

Government of Australia. (2007, March). *Review of implementation of the CBD (Convention on Biological Diversity) programme of work on protected areas* [National report]. Canberra, ACT: Government of Australia. Retrieved from www.cbd.int/doc/world/au/au-nr-ripa-en.doc

GER (Great Eastern Ranges Initiative). (2012). *The Great Eastern Ranges Initiative business plan 2012–2015.* Sydney, NSW: Great Eastern Ranges Initiative.

GER. (2015). *Great Eastern Ranges Initiative science plan 2015–2025.* Wollongong, NSW: Great Eastern Ranges Initiative.

GER. (2016). *Great Eastern Ranges corridor (draft) strategic plan 2016–2025*. Sydney, NSW: Great Eastern Ranges Initiative.

Great Eastern Ranges. (n.d.). *About us* [Great Eastern Ranges website]. Retrieved from www.greateasternranges.org.au

Howling, G. M. (2012). *Options for prioritising future NSW investment in the Great Eastern Ranges corridor*. Wollongong, NSW: Office of Environment and Heritage.

Howling, G. M. (2013). *Landscape investment plan for the Kanangra–Boyd to Wyangala link*. Wollongong, NSW: Office of Environment and Heritage.

Hyder Consulting. (2008). *The natural heritage values of the NSW section of the Great Eastern Ranges corridor*. Sydney, NSW: Hyder Consulting Pty Ltd.

IUCN (International Union for the Conservation of Nature) – World Commission on Protected Areas. (2006, 17 November). *The Papallacta declaration*. Retrieved from www.activeremedy.org/wp-content/uploads/2014/10/the-papallacta-declaration-2006.pdf

Mackey, B., Ferrier, S. & Possingham, H. P. (2013). Connectivity conservation principles for Australian wildlife corridors. In J. Fitzsimons, I. Pulsford & G. Wescott (Eds.), *Linking Australia's landscapes: Lessons and opportunities from large-scale conservation networks* (pp. 233–43). Melbourne, VIC: CSIRO Publishing.

Mackey, B. & Hugh, S. (2010). *Spatial analysis of conservation priorities in the Great Eastern Ranges, project 3: Productivity analysis and drought refuge*. Canberra, ACT: CSIRO Ecosystem Sciences.

Mackey, B., Watson, J. & Worboys, G. L. (2010). *Connectivity conservation and the Great Eastern Ranges corridor*. Retrieved from www.environment.nsw.gov.au/resources/nature/ccandger.pdf

McKinney, M., Scarlett, L. & Kemmis, D. (2010). *Large landscape conservation: A strategic framework for policy and action*. Cambridge, MA: Lincoln Institute for Land Policy.

National Parks and Wildlife Service. (2003). *The bioregions of New South Wales: Their biodiversity, conservation and history*. Hurstville, NSW: National Parks and Wildlife Service.

Office of Environment and Heritage. (2011). *Great Eastern Ranges Initiative: A continental-scale lifeline connecting people with nature*. Wollongong, NSW: Office of Environment and Heritage.

Pulsford, I., Worboys, G. L. & Howling, G. M. (2010). Australian Alps to Atherton connectivity conservation. In G. L. Worboys, W. L. Francis & M. Lockwood (Eds.), *Connectivity conservation management: A global guide* (pp. 96–105). London, UK: Earthscan.

Sattler, P. & Williams, R. (1999). *The conservation status of Queensland's bioregional ecosystems*. Brisbane, QLD: Environment Protection Agency, Queensland Government.

Smith, A. (2010). *Birds in the Great Eastern Ranges, movement and connectivity* (Report No. 2). Sydney, NSW: Department of Environment, Climate Change and Water.

Soulé, M. E. & Terborgh, J. (Eds.). (1999). *Continental conservation: Scientific foundations of regional reserve networks*. Washington, DC: Island Press.

Williams, K. J., Ford, A., Rosauer, D. F., De Silva, N., Mittermeier, R., Bruce, C., … Margules, C. (2011). Forests of east Australia: The 35th biodiversity hotspot. In F. E. Zachos & J. C. Habel (Eds.), *Biodiversity hotspots* (pp. 295–310). Berlin, GM: Springer-Verlag. doi.org/10.1007/978-3-642-20992-5_16

Worboys, G. L. (2010). The connectivity conservation imperative. In G. L Worboys, W. L. Francis & M. Lockwood (Eds.), *Connectivity conservation management: A global guide* (pp. 3–21). London, UK: Earthscan.

Worboys, G. L., Ament, R., Day, J. C., Locke, H., McClure, M., Tabor, G. & Woodley, S. (2016). *Connectivity conservation guidelines part one: Definition, area of connectivity conservation*. Gland, CH: International Union for The Conservation of Nature – World Commission on Protected Areas.

Worboys, G. L. & Pulsford, I. (2011). *Connectivity conservation in Australian landscapes*. Canberra, ACT: Australian Government Department of Sustainability, Environment, Water, Population and Communities.

13

Monitoring and Reporting Land Use Change and Its Effects on the Queensland Environment

Paul Lawrence, Craig Shephard, Phillip Norman, Christina Jones and Christian Witte

Key Points

- Land use datasets and their application to inform and influence decision-making require a combination of technology, scientific credibility, interpretative artistry and cross-disciplinary collaborations.

- The Queensland Land Use Mapping Program (QLUMP) has been instrumental in providing land use information to assist decision-making and investments by government to reduce the pollutant loads from catchments adjacent to the Great Barrier Reef.

- Products generated through QLUMP, such as the type of land use and monitoring changes in land use patterns across Queensland, provide lines of evidence to support priority programs, such as the Reef Water Quality Protection Plan, the South East Queensland Regional Plan, strategic cropping land and the State of the Environment.

- These applications demonstrate the value of adhering to the Australian Collaborative Land Use and Management Program (ACLUMP) guidelines to ensure a methodology is consistent, accurate, reliable, cost-effective and makes best use of available ancillary databases and data management infrastructure.

- Several emerging land use policy and planning issues in Queensland, and benefits from scientific and collaborative arrangements for land use mapping to inform and influence state and national scale issues are explored.

Introduction

Land use information is critical for planning, policies and decision-making. The availability of consistent and reliable information is essential for sustainable natural resource management (NRM) and environmental outcomes for local, state and federal governments, regional NRM groups, industry groups, community groups and land managers. The value of land use information is particularly evident when it is available at temporal and spatial scales that are fit for purpose, and when it is analysed in combination with other spatial datasets that inform decision-making, such as modelling, monitoring and economic evaluations.

The Queensland Spatial Information Council defined land use as a foundation spatial dataset that is 'vital for the progression and development of Queensland' (Queensland Government, 2017). The use of and reliance on land use mapping for priority programs and government initiatives have increased in recent years. The purpose of this chapter is to review some of the applications in Queensland that have benefited from the incorporation of land use information, both as a primary source and in secondary and supportive roles. Case studies, including the Reef Water Quality Protection Plan, the South East Queensland Regional Plan and Cape York Regional Plan, serve to highlight the use of land use information to assist decision-making and investments by government. This chapter shows the value of a nationally recognised standard in land use classification. This is particularly evident for cross-jurisdictional programs, such as the Murray–Darling Basin and agricultural development in northern Australia. Observations on the future directions and technological challenges for land use mapping for dynamic reporting—particularly the extent to which these might facilitate the continuing enhancement of land use planning products, readying them to inform complex non-routine and multi-stakeholder NRM issues—are offered by way of conclusion.

Land Use Mapping in Queensland

The Queensland Land Use Mapping Program (QLUMP) is the primary vehicle for deriving spatial and change detection information on land use within the state. The program is a partnership between the Remote Sensing Centre, Department of Science, Information Technology and Innovation (DSITI) and Regional Service Delivery, Department of Natural Resources and Mines. The input of regional staff throughout Queensland is crucial; their mapping skills, local knowledge and capacity to engage regional experts in compiling updated land use mapping contribute to the overall accuracy of the program.

Since commencing in 1998, QLUMP has been active within the Australian Collaborative Land Use and Management Program (ACLUMP), which includes all jurisdictions and is coordinated by the Australian Government's Department of Agriculture and Water Resources (Australian Bureau of Agricultural and Resource Economics and Sciences, 2011). Mapping is undertaken to a national standard, and Queensland scientists have influenced the national mapping methodology; this reflects the adaptive and applied nature of the land use mapping framework, as well as the success of the collaborative partnership.

The technical foundation of QLUMP has evolved over time; it utilises the latest techniques for identification, compilation and management of spatial data. Originally, the land use maps were compiled in a raster environment, using Landsat TM (Thematic Mapper) and ETM+ (Enhanced Thematic Mapper Plus) satellite imagery. Improvements in vector data enabled QLUMP to transition to editing in a vector environment (e.g. the Earth Sciences and Resources Institute's ArcGIS, a geographic information system [GIS] for working with maps). More recent applications, including the 2009 maps showing land use in the Great Barrier Reef catchments, have benefited from innovative workflow systems to manage and coordinate land use mapping by individual spatial officers across the state. For example, the Earth Sciences and Resources Institute's ArcSDE (Spatial Database Engine) geodatabase replication infrastructure efficiently allocated each spatial officer a specific region requiring update; provided spatial officers with the most up-to-date version of the data; enabled spatial officers to map their region on their own computer before submitting edits to the original database; and authorised QLUMP managers to perform quality assurance, before committing data to the original database. Through this process, over 800,000 individual edits were performed by QLUMP across nine regional offices in Queensland, with each edit checked within a quality-assured framework. Without the efficient management and exchange of spatial data, the update of land use in the Reef catchments—which cover 380,000 square kilometres—to a consistent format, could not have been achieved within the time frame.

Further developments in QLUMP methods have been driven, in part, by improved ancillary data; in particular, the increasing availability of suitable datasets that aid land use interpretation. These have been incorporated into workflow processing through a GIS dichotomous decision-tree approach (Lawrence & Shephard, 2014). Ancillary data layers are queried in accordance with the Australian Land Use and Management Classification (ALUMC) hierarchy (in combination with the decision rules) to output a 'flattened' spatial layer representing land use; in certain circumstances, known features (e.g. estates) can be 'cut' straight into the mapping layer to aid efficiency and accuracy.

QLUMP now utilises a tablet personal computer to undertake field-based editing, run ArcGIS and access all the ancillary and imagery data normally available at the desktop. This allows officers to efficiently edit and annotate land use maps in the field, thereby reducing the duplication of work.

Advances in the acquisition, access, availability, resolution, cost and timeliness of suitable imagery have greatly influenced the quality of land use data. Whereas image data were once scarce and coarse (e.g. Landsat 30 metres), QLUMP now uses high-resolution state-wide imagery (e.g. SPOT 6/7 1.5 metres) to update land use, and higher-resolution orthophotography (10 centimetres) is available for some coastal catchments. In addition, there are numerous freely available image sources, such as Google Earth and Street View imagery, that have proved to be great resources for updating land use in intensive regions (e.g. south-east Queensland).

While higher spatial resolution imagery is becoming more accessible, the challenge for land use mapping remains the temporal resolution. This presents limitations for mapping and decision-making, particularly for seasonal or opportunistic land uses (e.g. summer–winter cropping) and temporary or episodic events (e.g. flood mapping, channel erosion and fire). Typically, the acquisition of highly temporal and coarse data (e.g. Landsat 30 metres) is prompt and free, while the acquisition of high-resolution data (e.g. orthophotography) is delayed and expensive. Improving the supply chain and delivery of high-resolution imagery would result in significant improvements for providing timely land use information. The spatial resolution of imagery is also important with respect to the intensity of land use. The greatest efficiencies are found in applying a mix of both high-temporal (yet course) resolution imagery and high-spatial (yet untimely) resolution imagery to compile the most recent land use maps.

A recent example of this approach is the updated land use map for the Tully catchment in north Queensland. In response to the recent Panama Disease Tropical Race 4 biosecurity issue, QLUMP acquired the most recent SPOT 6/7 satellite imagery for the agricultural production areas within the Tully catchment, supplemented by coarser imagery elsewhere. This enabled QLUMP to compile a five-week-old high-resolution land use map for Biosecurity Queensland. Information management and knowledge exchange remain core pillars of the QLUMP system. The QLUMP DocuWiki environment (i.e. in-house documentation) ensures a consistent approach to the management of work procedures, nomenclature in file naming, data sources, decision rules, terminology and access to knowledge.

Some QLUMP Applications

Land use mapping (see Figure 13.1) allows governments, stakeholders and land managers to:

- describe the type and extent of land uses, and explain how land use decisions align with policy goals and their environmental, economic and social challenges
- support analysis of the extent and effects of land uses on agriculture, natural resources, the environment and regional communities
- consider the influences of land use change decisions relative to on-site and off-site impacts, and assess changes in land use that result from regulation, policy and incentives.

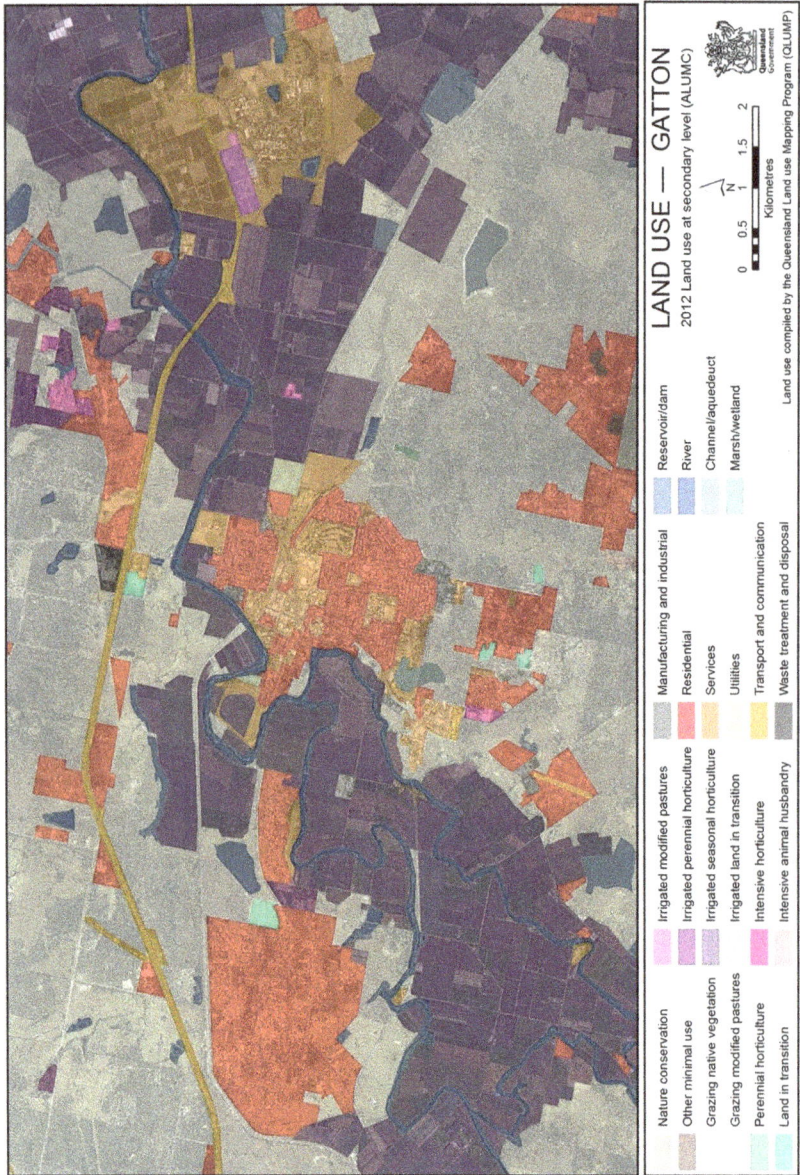

Figure 13.1: Land use mapping example—Gatton, South East Queensland.

Source: Queensland Government (2017).

The demand for timely land use information is broad; it is applied across a range of NRM issues, with particular focus on the rural sector. During 2013, there were approximately 1,700 downloads of QLUMP datasets from the Queensland Spatial Catalogue—QSpatial. The portfolio of applications and supplementary mapping covers a spectrum of information and products, including:

- agricultural productivity and sustainability (i.e. profitable production of food and fibre and adoption of sustainable agricultural practices)
- land use planning (i.e. supporting regional planning and investment, and strategies for development)
- biosecurity (i.e. managing invasive species and minimising the impact of incursions, managing weeds and feral animals and their impact on threatened species and evaluating the risk of disease spread in crops)
- natural resource condition monitoring and investment (i.e. setting soundly based targets and monitoring procedures for natural resource investment at national, state, regional and local levels of responsibility)
- biodiversity conservation (i.e. managing and mitigating the effect of production systems on terrestrial, aquatic, coastal and marine habitats)
- improving water availability and quality (i.e. responding to water allocation and efficiency needs; responding to water deficits arising from drought and the need for increased environmental flows; and managing water quality, including sediment and nutrient loads)
- natural disaster management (i.e. preparing for, responding to and evaluating the impact of events such as floods, cyclones, bushfires and drought).

In updating catchment-scale land use mapping, QLUMP also revises older mapping to account for improvements or corrections. Defensible land use change data that show *actual* land use change, rather than improvements to the mapping, are then derived at the secondary level of the ALUMC to reflect the consistency of mapping across the catchment. For example, land use change from rural residential to urban will not appear in the land use change dataset; this is because it is at the tertiary level of the classification.

QLUMP produces land use summary reports for each catchment, including maps, summary statistics, data limitations and results of the accuracy assessment. Land use change is presented relative to the change in intensity of the land use at the secondary level of the ALUMC.

A change from 2.1.0 (grazing native vegetation) to 2.2.0 (production forestry) represents an increase in land use intensity, while a change from 2.1.0 (grazing native vegetation) to 1.1.0 (nature conservation) represents a decrease. Further analysis of the change, both from and to specific land use classes, is also undertaken. As a result, change can be expressed in terms of extent (in hectares or percentages) or intensification–deintensification. An experimental weighted change model is being considered to reflect the range of intensity of land use changes and account for the 'absorbing state'. For example, a land use change from one estate (e.g. managed resource protection) to another (e.g. nature conservation) is not as significant as a land use change to an intensive land use class (e.g. mining or residential). However, further value-adding with catchment modelling or land use pattern modelling remains largely untested.

Another priority application for QLUMP information is the parameterisation of catchment-scale modelling for the Reef Plan. The alignment of land use with management practice data is essential for accuracy in delivering credible responses towards achieving water quality and management targets on an annual basis. Changes in land use within the reef catchments are now routinely incorporated into the recalibrated catchment modelling. Similarly, the crop frequency model, which informs Queensland's *Strategic Cropping Land Act 2011*, relies on QLUMP data to reduce and mitigate commission errors.

Critical Issues and Future Opportunities

The full potential of land use mapping to inform decision-making is constrained by several critical issues.

Currency

While information for parts of Queensland has been updated, the 1999 (state-wide) baseline is still the only data available for 50 per cent of Queensland (see Figure 13.2). QLUMP has progressively updated catchment-scale land uses on an ad hoc basis, generally in response to policy demands and acquisition of suitable imagery. For instance, the Great Barrier Reef catchments data were updated in 2009 to support the Reef Plan. Ideally, QLUMP strives to maintain the currency of land use in Queensland as per ACLUMP guidelines (i.e. nominally five years in coastal catchments and 10 years elsewhere).

QLUMP Currency

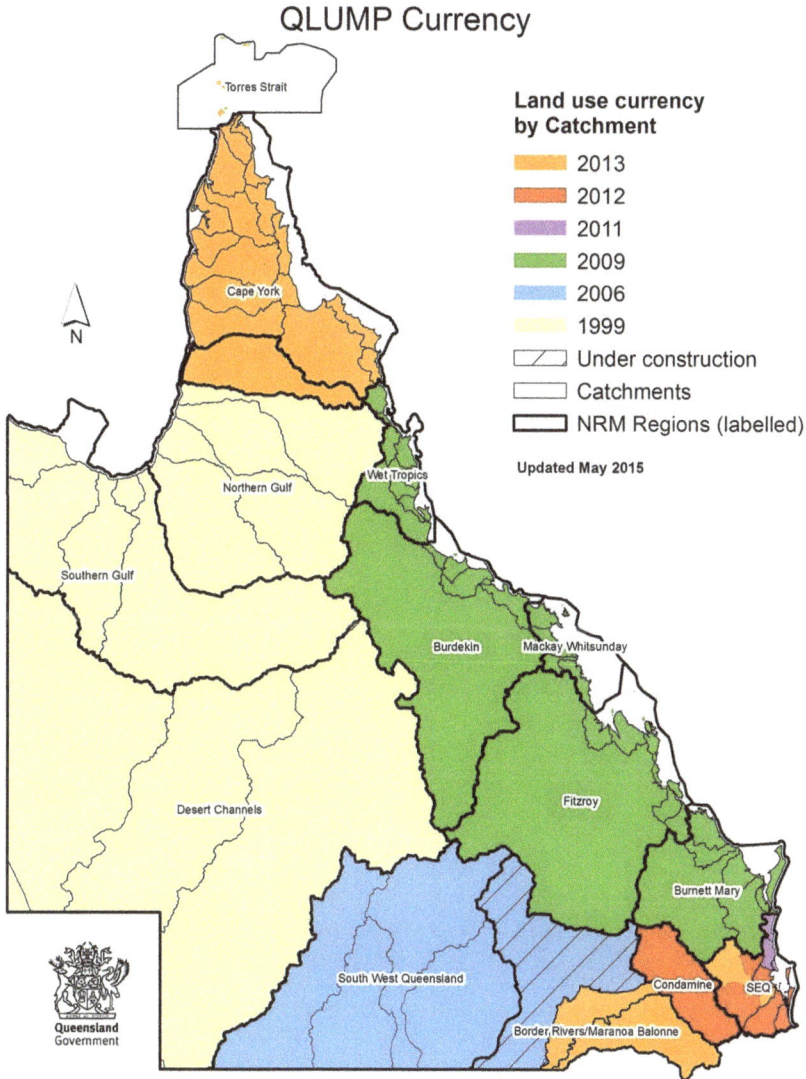

Figure 13.2: Land use currency mapping for Queensland.
Source: Queensland Government (2017).

Imagery

QLUMP relies on the government's Spatial Imagery Acquisition Program (SIAP) to acquire suitable imagery. Negotiated licence agreements provide access to contributing agencies. While the main focus of the program is aerial photography captured for urban and intensive land use areas, high-resolution state-wide satellite imagery purchases have also been supported—specifically, 2009 and 2012 SPOT 5 imagery, and 2013 SPOT 6 imagery. Suitable imagery for western regions and catchments (e.g. Landsat) is available without cost through the United States Geological Service – NASA website, while SPOT 5 (or similar) satellite imagery and high-resolution orthophotography are purchased for specific regional priorities (e.g. Murray–Darling Basin and reef catchments) (DSITI, 2015). The QLUMP program works closely with SIAP to drive potential efficiencies, as QLUMP is well positioned to update and add value to the land use component (from all suitable imagery) for other spatial mapping products. Targeting potential regional growth areas through a compilation of spatial information (including land use type and change) should also be considered to improve the full return on investment.

Data Substitution

QLUMP has participated in ACLUMP pilots that seek to improve the currency of existing land use data through alternative sourcing. One such pilot, Updating Land Use by Exception, used land valuation data from the Queensland Valuation System (QVAS) to compile land use and land use change (Ground, Lawrence & Shephard, 2013). The success of this approach revealed several limitations in terms of spatial and temporal accuracies; for example, it returned a low correlation (Kappa) score. Figure 13.3 shows an example of the differences in land use mapping for a section within the Goondiwindi region of the St George area, highlighting the differences in type and extent of land use change between 1999 and 2006. These differences may be attributed to the nature of QVAS data, which are not updated to the same frequency or consistency. Additionally, QVAS is geometrically derived from the Digital Cadastral Database (DCDB) and cannot represent land use at the sub-parcel level. While some jurisdictions have opted to use valuation property classification coding, this approach is not recommended for Queensland, where spatial and thematic accuracy at the sub-parcel level provides the best support for science, policy and planning applications.

Figure 13.3: Comparison of land use mapping classification at sub-parcel level—Goondiwindi region, south-west Queensland.

Source: Queensland Government (2017).

Impact Mapping

The spatial change in land use may not be proportional to the impacts on water quality and ecosystem services. Although extent and resolution are fundamental, impacts from relatively small footprint industries,

213

such as coal seam gas installations, vegetation within ephemeral macro-channels or land management practices, may not be detected from coarser resolution imagery, yet these may contribute to catchment processes and environmental management. Consequently, a challenge for remote-sensing sciences is to utilise spatial imagery across a range of spatial and temporal scales to detect micro-scale, high-impact changes.

Recommendations

Land use datasets are key economic enablers for the development of the state's natural resources; at the same time, they maintain a high degree of environmental protection by objective measurement. The data are used widely across the Queensland Government, and externally by regional NRM bodies, industry bodies, consultants, local governments, research agencies and community groups. The national coordination provided by ACLUMP has been critical to the success of land use mapping in Queensland. The consistent and defensible mapping generates information, knowledge and communication products to inform policy and planning decisions with confidence, now and into the future.

Basic land use types are now routinely monitored and mapped in Queensland; however, access to new imagery to provide increased temporal and spatial resolution of land use is required to mature these early gains in land use identification. New concepts that involve heuristic, multiple-criteria analysis and dichotomous 'intelligent' identification of land use management must be researched and validated to respond to more complex and 'wicked' problems involving soil and water quality outcomes. Equally, changes in land use should be expressed in units that convey impact as well as extent of the mapping area.

References

Australian Bureau of Agricultural and Resource Economics and Sciences. (2011). *Guidelines for land use mapping in Australia: Principles, procedures and definitions* (4th ed.). Canberra, ACT: Australian Bureau of Agricultural and Resource Economics and Sciences.

DSITI (Department of Science, Information Technology and Innovation). (2015). *Land use summary 1999–2013: Border Rivers and Moonie catchments*. Brisbane, QLD: Department of Science, Information Technology and Innovation.

Ground, S., Lawrence, L. & Shephard, C. (2013, April). *Monitoring land use: Land use change and land management practices within the Queensland Murray–Darling Basin: A comparison of Queensland land use mapping programs and Queensland valuation systems*. ACLUMP Technical Workshop, Canberra, ACT.

Lawrence, L. & Shephard, C. (2014, April). *Updating land use mapping by exception—a GIS decision tree approach*. ACLUMP Technical Workshop, Bendigo, VIC.

Queensland Government. (2017). *Queensland land use mapping program (QLUMP)*. Retrieved from www.qld.gov.au/environment/land/vegetation/mapping/qlump-background

14

Down Payments on National Environmental Accounts

Michael Vardon, John Ovington, Valdis Juskevics,
John Purcell and Mark Eigenraam

Key Points

- Environmental accounting is promoted in Australia and around the world as a way of providing regular information on changes in the state of the environment to support policy development and decision-making.

- In Australia, significant progress on the development and implementation of environmental accounting has been made at different levels of government, academia, corporations and non-government organisations. This includes the production of accounts for mineral and timber assets, water, waste, land and ecosystems.

- Australia's efforts represent a significant down payment on national environmental accounting. Standards have been recommended and a range of agencies have experience in the production of the accounts.

- However, there is a lack of knowledge of environmental accounting at senior levels of government, and how it can be applied to mainstream policy- and decision-making, and this is a major barrier.

- Attaining the resources and institutional arrangements needed to regularly produce accounts and use them in decision-making processes will require a greater focus on meeting the needs of decision-makers.

- In memory of Dr Rob Lesslie, his contributions to the development of environmental accounting, and land accounts in particular, are highlighted.

Introduction

Environmental accounting in Australia has been advanced by many organisations, including national and state government agencies, regional natural resource management (NRM) authorities, and business and non-government organisations. In addition, there has been a range of closely related activity on the development of information systems and clarification of concepts that are measured and reported in the accounts, such as ecosystem services and environmental condition.

A call for the development of environmental accounts has been made in several government documents relating to the environment (see Hawke, 2009; Morton & Tinney, 2011); it featured in a former government's vision of *Australia in the Asian Century* (Commonwealth of Australia, 2012) and is a specific focus of the National Plan for Environmental Information Initiative (Department of Environment and Energy [DEE], n.d.-a). Yet, despite these calls, progress has been limited. Key barriers to the development and use of environmental accounting have been (and remain) a lack of understanding about what environmental accounting is, and how environmental accounts can inform policies, programs and decisions.

What is Environmental Accounting?

At its simplest, environmental accounting is a way of organising information. It is based on a model of stocks and flows and the measurement of transactions between parties (e.g. a buyer and seller). For example, someone buying a bag of apples from a market exchanges money: $5 for a 4-kilogram bag. The buyer's stock of apples goes from 0 to 4 kilograms, while their stock of money goes down. If they had $100, the buyer's remaining stock of money would be $95. Meanwhile, the seller's stock of apples is reduced by 4 kilograms, and their stock of money is increased by $5. This type of accounting allows businesses to manage inventories, cash flows and assets. Business accounting of this type has been around for centuries and was formalised by Pacioli in 1496 (Gleeson-White, 2014).

National accounting for the purposes of managing entire economies, and not just business or government finances, evolved in the twentieth century. It grew out of the need to better manage the economy during the Great Depression and World War II (Obst & Vardon, 2014). Recognised shortcomings in the treatment of the environment in the national accounts eventually led to the creation of the System of Environmental-Economic Accounting (SEEA), which provides a means of recording the transactions between the environment and the economy. The SEEA was adopted by the United Nations (UN) as an international standard in 2012 (UN, 2014a)[1] and extended to cover ecosystem accounting in 2013 (UN, 2014b).[2]

The SEEA provides a system for integrating environmental information across domains (e.g. water, energy, pollution and biodiversity) with economic information. Physical information on the environment can be compared with economic transactions to help identify areas (e.g. coastal areas) or industries (e.g, agricultural) for particular attention based on the impact they are having on the environment, as well as the risk to economic activity from environmental change.

A common misconception of environmental accounting is that its aim is to measure and value everything in monetary terms. While it is true that the scope of environmental accounts includes both physical and monetary measures, these can be produced separately. Indeed, many accounts (and most of those produced so far in Australia) only include physical measures, as discussed below.

Environmental accounting has many potential applications. An independent review of the *Environmental Protection and Biodiversity Conservation Act 1999* (Hawke, 2009) listed numerous expected benefits from the production of regular environmental accounts, including:

---

1   The SEEA Central Framework was adopted at the 2012 meeting of the UN Statistics Commission; however, the final version was not published until 2014.

2   The SEEA Experimental Ecosystem Accounting was recognised at the 2013 meeting of the UN Statistics Commission; however, the final version was not published until 2014.

- providing measurable ways of comparing and assessing environmental assets over time
- providing a practical base for investing in future actions for environmental assets
- providing information to underpin evidence-based decision-making
- better targeting of private and public investment at the program and project level
- better measurement and understanding of the impacts and effectiveness of policies and investments
- allowing for better identification and management of risks
- providing greater community visibility on environmental outcomes
- guiding environmental and land use planning, including through environmental impact assessments and regional planning
- identifying and addressing gaps in reporting requirements and informing the State of the Environment reporting process.

Applications and uses of environmental accounting around the world have been summarised by Smith (2014) and the European Commission (2014). Use of water accounts in Australia were highlighted by Vardon, Lenzen, Peevor and Creaser (2007). The uses identified ranged from sophisticated input–output, or computable general equilibrium modelling, to the identification of trends in resource use, and a comparison of these to changes in the size of the economy or population (i.e. the so-called decoupling indicators). The application of accounts to real-world management and policy is an area in need of more detailed exploration (Vardon, Burnett & Dovers, 2016).

Environmental Accounting in Australia

An overview of environmental accounting in Australia is summarised in Figure 14.1. The Australian Bureau of Statistics (ABS) has been involved in the development of environmental accounting for over two decades and has produced the largest number of accounts.

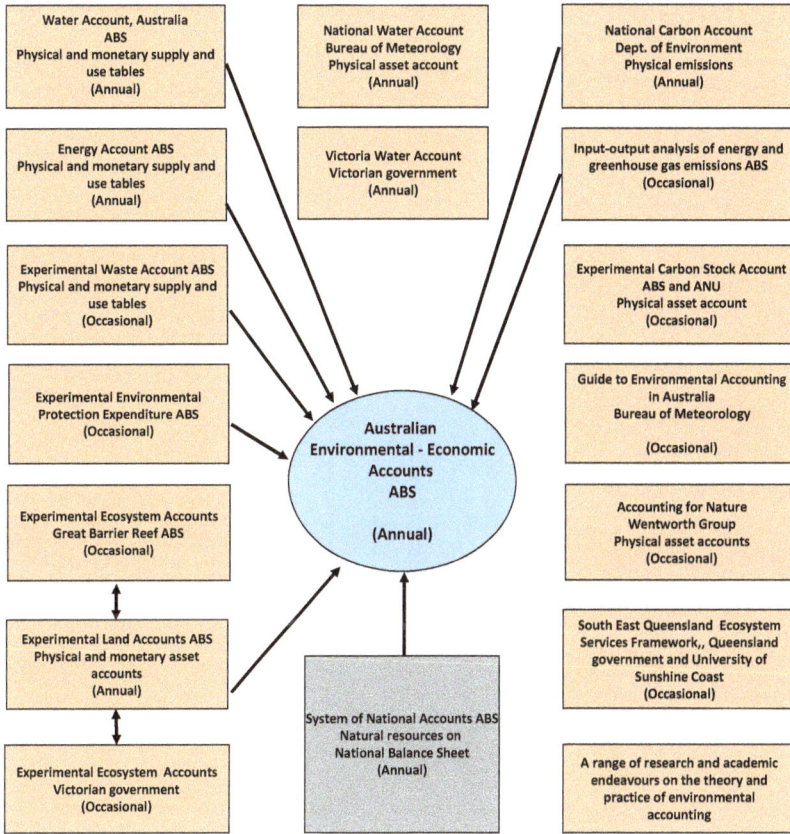

Figure 14.1: Overview of environmental accounting in Australia.
Source: Australian Bureau of Statistics. Used with permission.

Table 14.1 provides further details of the ABS accounts, including the year first published, time series available and coverage in terms of stocks and flows and monetary and physical measures. The Australian Environmental-Economic Accounts (AEEA) draws together all ABS accounts; the second edition was published in 2015 (ABS, 2015a). Other major environmental accounting activity by government include the National Greenhouse Gas and Energy Reporting (DEE, n.d.-b) and the National Water Account (Bureau of Meteorology [BoM], 2014). All accounts have improved over time with better understanding of the data used in the compilation of accounts, identification of additional data, collection of new data and improvements in the systems and processes used to produce accounts.

Table 14.1: Environmental-economic accounts produced by the Australian Bureau of Statistics.

Account type	Year first published	Frequency or status	Reference years for which accounts are available			
			Stock accounts		Flow accounts	
			Physical	Monetary	Physical	Monetary
National balance sheet	1995	Annual from 1995	1988–89 to 2012–13	1988–89 to 2012–13		
-Land						
-Minerals						
-Energy						
-Timber						
-Fish	2012	Experimental		2000–01 2005–06 to 2009–10		
Fish	1995	Occasional	1996–97		1996–97	
Energy	1996	Annual from 2011	1988–89 to 2011–12	1988–89 to 2011–12	2008–09 to 2012–13; 2004–05, 2006–07, 1993–94 to 1996–97	2011–12, 2009–10, 2004–05
Minerals	1998	Occasional	1985–1996		1992–93, 1993–94	
Water	2000	Annual from 2010			2008–09 to 2012–13, 2004–05, 2000–01, 1993–94 to 1996–97	2008–09 to 2012–13; 2004–05; 2003–04
Land cover and land use values (by state)	2011	Annual from 2011	2011–2014	2011–2014	2011–2014	2011–2014
Waste	2012	Discontinued			1997–98 to 2011–12	2009–10
GHG emissions embedded in final demand	2012	Experimental			2008–09	

Note: Land cover and use accounts are prepared for each state on a rotating schedule, with each state covered once in three years.

Source: Obst & Vardon (2014).

Many academics have been involved in research relating to ecosystem accounting (e.g. Aisbett & Kragt, 2010; Binning, Cork, Parry & Shelton, 2001; Crossman et al., 2013; Gillespie, Dumsday & Bennett, 2008; Pittock, Cork & Maynard, 2012; Russell-Smith et al., 2013; Stoeckl et al., 2011; Straton & Zander, 2009; Tovey, 2008; van Dijk et al., 2014). This activity has added significantly to the theory and practice of environmental accounting in Australia.

Different initiatives have focused on various aspects of the accounts in terms of concepts, themes or metrics. For example, ecosystem service flows were the focus of the Department of Environment, Water, Heritage and Arts (2009); Maynard, James and Davidson (2010); and Cork et al. (2012). However, the condition of ecosystems assets was the focus of the Wentworth Group of Concerned Scientists (2008, 2013; Sbrocchi 2015). The ABS (see Table 14.1) compiled accounts in both monetary and physical metrics for land, water, waste and energy using information on stocks and flows; by contrast, water (in physical terms) has been the main focus of the BoM.

There has been some confusion about the water accounts published by the ABS and BoM. It is important to recognise that water accounting involves a series of accounts; the SEEA-Water describes 12 different types of standard accounts (UN, 2012)—less than half of which have been produced for Australia. The main differences are that the ABS Water Account, produced since 2000, explicitly follows the SEEA and is composed of physical and monetary supply use tables. The BoM National Water Account, produced since 2007, is primarily an asset account in physical terms. The BoM account follows standards developed by the Water Accounting Standards Board and does not explicitly use the SEEA, although it can be mapped into the SEEA (Vardon, 2012).

The ABS and BoM have also worked closely together on the development of accounting frameworks. In 2013, the BoM recommended the adoption of the SEEA (BoM, 2013). Together, these two agencies coordinated the Australian inputs to the drafting of the System of Environmental-Economic Accounting – Experimental Ecosystem Accounting (SEEA-EEA) (UN, 2014b). As an example of this coordination, in May 2012, the ABS, BoM and Victorian Department of Sustainability and Environment hosted an Expert Meeting on Ecosystem Accounts (2012) as part of the SEEA-EEA development.

There have also been environmental accounting initiatives by the corporate sector. Business accounting for the environment has proceeded along a variety of fronts (see Gleeson-White, 2014). This type of accounting is concerned with the management of individual businesses, rather than the environment or economy as a whole. While the basic principles of this type of accounting are similar, and much can be learned from the approaches used by businesses—in particular, how accounts are used in management—it is not given further consideration here, as the focus is on national environmental accounts.

Land Accounting

This section examines the development of land accounting in Australia. The reason for the focus on the development of these types of accounts is fourfold:

1. Land accounting is the foundation of ecosystem accounting, which is at the leading edge of accounting nationally and internationally.
2. Land accounting provides a good example of the theoretical and practical development of information sources and accounts in Australia.
3. There are existing summaries of water (see Godfrey & Chalmers, 2012) and greenhouse gas reporting (see DEE, n.d.-c), which are the other accounts in regular production.
4. In memory of Rob Lesslie, who was involved in the development of land accounts from the beginning.

Rob Lesslie was a key contributor to land accounting in the early stages of methodological development and provided vital support and encouragement to the ABS and others involved. He also provided a link to earlier work on land information (e.g. National Land and Water Resources Audit) and important input to measuring vegetation condition at large scales via the development of the vegetation assets, states and transitions (VAST) framework (Thackway & Lesslie, 2006).

The key to the integration of environmental and economic information via accounts is spatially explicit data. This can be illustrated by the development of integrated environmental-economic accounts for land by the ABS. This has been prepared for several states (i.e. Victoria, South Australia and New South Wales [NSW]), the Great Barrier Reef Region and the Murray–Darling Basin. The Great Barrier Reef is an area of international significance; it is listed on the UN Educational, Scientific, Cultural Organization's World Heritage List. The Murray–Darling Basin is an important agricultural area that is under increasing environmental and economic pressure from the limited availability of water, which is exacerbated by droughts. The focus here is on the Great Barrier Reef, for which both land and ecosystem accounts have been completed (ABS, 2011, 2014b, 2015b).

In February 2011, the ABS released the first experimental land accounts for the Great Barrier Reef region. These accounts were the culmination of seven years of investigative work; they were followed by a second set of land accounts (ABS, 2014b) and ecosystem accounts (ABS, 2015b). The accounting work grew out of surveys of land management practices that were initially conducted in NSW and Queensland. The surveys were published as:

- Eurobodalla (NSW) (ABS, 2005)
- Fitzroy and Livingstone Shires (Queensland) (ABS, 2006)
- Great Barrier Reef catchments (Queensland) (ABS, 2010).

These surveys embodied several innovations. In Eurobodalla, the land parcel, comprising a defined spatial area, was used as the statistical unit of observation, rather than the traditional business unit (i.e. the farm business). The use of a land parcel as the statistical unit provided the opportunity to explicitly link the biophysical information available (e.g. from remote sensing) to the management practices on the land parcel and other administrative data. The initial work identified several areas for methodological improvement; it also revealed opportunities, such as how other information sources could be incorporated into the approach.

The studies of the Fitzroy and Livingstone Shires developed a new approach for collecting land-based data. The survey design was based on the cadastre (land boundaries based on ownership). This was combined with administrative data held by state and local governments, enabling a random selection of land parcels of agricultural land to which ownership details could be attached. The survey was the first to use forms tailored to each selected business. Each survey included a map that showed every cadastral parcel owned by that business; owners were asked to provide information for the selected parcels only. An unintended but important and welcome outcome of this strategy was that it encouraged landowners to become more engaged with the survey, resulting in improved response rates and improved data quality.

The flexibility of using a spatial unit as the primary statistical unit was apparent in the information presented for the Fitzroy and Livingstone Shires (ABS, 2005). In this, the data could be aggregated to different geographies: by shire, the riparian zone of the Fitzroy River, the coastal zone, by radial zones from the Rockhampton City Centre and by 23 neighbourhood catchments. In a traditional business survey approach, only one set of geographical boundaries would have been possible (i.e. by shire).

The 2009 survey of land management practices of the Great Barrier Reef catchments further refined the land parcel methodology. Conducted under the auspices of the Australian Government's Reef Rescue initiative, it provided information on the land management practices affecting the amount of sediment reaching the Great Barrier Reef. Financial support was provided via the Caring for our Country initiative. Other support came from the Queensland Government via the Department of Environment and Resource Management, and from the office of the Valuer-General, which provided access to cadastral data.

This work led to the publication of the first experimental land accounts (ABS, 2011). These accounts presented information on land use, land cover and the value of land. They included data from a wide range of sources, such as the:

- Public Sector Mapping Agency (for the provision of the Cadlite® and Transport and TopographyTM)
- BoM (climate data)
- Geoscience Australia and the Australian Bureau of Agricultural and Resource Economics (dynamic land cover mapping)
- Department of Climate Change and Energy Efficiency (forest extent and change)
- Australian Bureau of Agricultural and Resource Economics (catchment-scale land use of Australia)
- North Australian Fire Information (fire frequency data)
- Department of Environment and Resource Management (land use classifications and land values for land parcels; data on wetlands)
- Department of Sustainability, Environment, Water, Population and Communities (wetlands data).

Land accounts for the Great Barrier Reef region were compiled and published by the ABS in 2014. Again, a wide variety of data sources were used; however, in this case, the ABS collected no additional data (i.e. the information presented was compiled from existing data sources within and outside the ABS). The production of the second set of accounts was important, as it demonstrated that repeat production was possible. It also meant that changes between the two time periods could be presented (see Figures 14.2a and 14.2b).

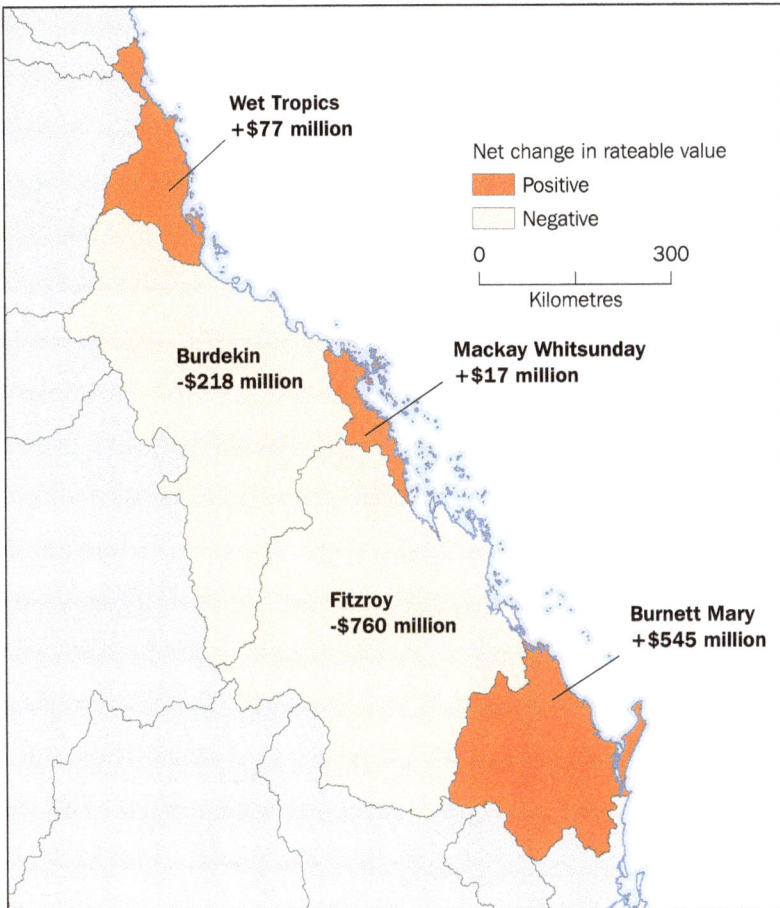

Figure 14.2a: Net change in rateable value of land used for livestock grazing between 2009 and 2013.

Source: ABS (2014a).

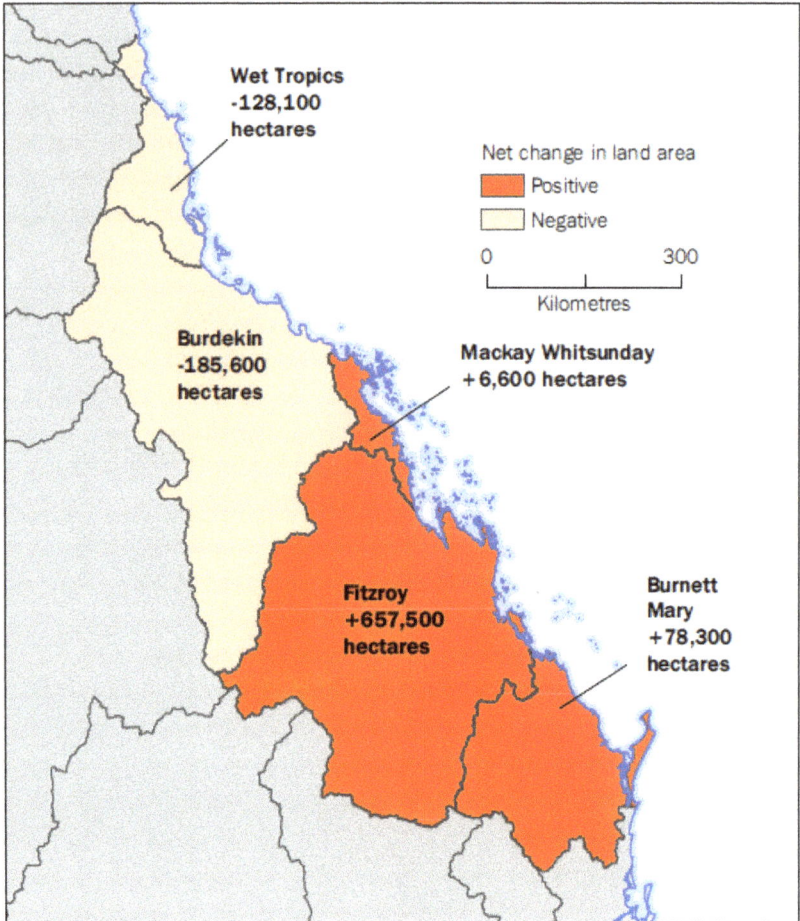

Figure 14.2b: Net change in land used for grazing 2009–13.
Source: Compiled from existing data sources within and outside the Australian Bureau of Statistics.

Land accounting is now part of the regular production of environmental accounts by the ABS. A land cover account based on dynamic land cover data is produced annually as part of the AEEA, while the total value of land is recorded on the national balance sheet. State land use accounts are scheduled to be produced every three years. So far, land accounts have been produced for Victoria (in 2012) and Queensland (ABS, 2013).

Ecosystem Accounting

Ecosystem accounting, focusing on the role of natural capital and its ecosystem services, has been an area of interest in Australia for more than a decade (e.g. Aisbett & Kragt, 2010; Binning et al., 2001; Crossman et al., 2013; Gillespie et al., 2008; Pittock et al., 2012; Russell-Smith et al., 2013; Straton & Zander, 2009; Tovey, 2008). The independent think tank Australia21 prepared a discussion paper on ecosystem services for the Department of Agriculture, Fisheries and Forestry that examined the literature relating to ecosystem services from Australia and around the world (Cork et al., 2012). The paper considered how ecosystem services could contribute to policy and management in relation to natural resources and human wellbeing in Australia. It concluded that there are issues to be addressed in relation to how an ecosystem services approach might be put into practice. At the time, a range of additional activity was in progress; since then, examples of how the concepts of ecosystem services and natural capital can be put into practice have emerged. These are summarised below.

Victoria

The Victorian Government was the first in Australia to produce ecosystem accounts (see Eigenraam, Chua & Hasker, 2013). They were linked directly to the ABS land accounts (ABS, 2012a) and used the same spatial output regions. Victoria is an interesting example because its accounts evolved along with the administrative processes that both used and populated the accounts. Further, they grew out of market-based schemes designed to promote the retention or regeneration of native vegetation on private land (see Stoneham, O'Keefe, Eigenraam & Bain, 2012).

The Victorian ecosystem accounts show changes in native vegetation area and condition by type of native vegetation between 1750 and 2005 (see Table 14.2). Similar tables are available for each NRM region, bioregion and statistical area. The level of change varied between vegetation types; eucalypt woodlands suffered the largest absolute fall in area (nearly 6 million hectares), while callitris forests, woodlands and tussock grasslands shared the largest fall in condition, from 1.00 to 0.33 habitat hectares (along with a large percentage of change in terms of area).

Table 14.2: Victorian terrestrial extent and condition classified by major vegetation groups: 1750 and 2005.

Major Vegetation Group (NVIS)	1750 Extent (Ha)	1750 Mean condition/Ha	2005 Extent (Ha)	2005 Mean condition/Ha
Native vegetation				
Acacia Forests and Woodlands	22,885	1.00	41,237	0.60
Acacia Open Woodlands	271	1.00	NA	NA
Acacia Shrublands	15,874	1.00	109	0.35
Callitris Forests and Woodlands	5,549	1.00	464	0.33
Casuarina Forests and Woodlands	1,003,122	1.00	186,411	0.48
Chenopod Shrublands, Samphire Shrublands and Forblands	214,488	1.00	55,516	0.51
Eucalypt Open Forests	6,346,166	1.00	3,899,116	0.65
Eucalypt Open Woodlands	1,223,235	1.00	1	0.60
Eucalypt Tall Open Forests	53,605	1.00	632,333	0.68
Eucalypt Woodlands	7,532,842	1.00	1,559,369	0.57
Heathlands	299,343	1.00	35,914	0.63
Low Closed Forests and Tall Closed Shrublands	206,330	1.00	NA	NA
Mallee Open Woodlands and Sparse Mallee Shrublands	213,785	1.00	NA	NA
Mallee Woodlands and Shrublands	3,395,152	1.00	1,509,023	0.56
Mangroves	7,025	1.00	1,010	0.53
Melaleuca Forests and Woodlands	89	1.00	14,910	0.50
Naturally bare - sand, rock, claypan, mudflat	4,619	1.00	3,066	0.35
Other Forests and Woodlands	63,290	1.00	287,940	0.59
Other Grasslands, Herblands, Sedgelands and Rushlands	202,082	1.00	142,010	0.59
Other Open Woodlands	122	1.00	NA	NA
Other Shrublands	295,419	1.00	103,193	0.61
Rainforests and Vine Thickets	44,109	1.00	36,630	0.71
Tussock Grasslands	1,302,356	1.00	28,486	0.33
Unclassified native vegetation	45,808	1.00	1	0.73
Total native vegetation	22,497,566	1.00	8,536,739	0.61

Source: Eigenraam et al. (2013, Table 1.1.0).

This account can be used to target particular vegetation types for attention in terms of either extent or quality, as well as the services they provide, including habitat for rare and threatened species. The accounts allow for structured comparisons within and between the regions and vegetation types in both absolute and relative terms (e.g. conversion to a percentage). The linking of these accounts to the economic activity by land use accounts highlights both the drivers of change and the benefits derived from change (e.g. agriculture, forestry, rural residential, etc.).

The quantification of both the area and environmental benefits resulting from government investments on private land (i.e. value for money)— as well as the cost of increasing these benefits—is a clear benefit of the accounts. For example, the result of one scheme in West Gippsland showed that for the $2.4 million invested, government achieved an extra 1,263 hectares of private land for conservation use, which represented an average cost of $380 per hectare per year (Eigenraam et al., 2013).

Part of the accounting done in Victoria involved quantifying the benefits expected in the future, and accounting for them now. An Environment Benefit Index (EBI) was calculated for all agricultural land in West Gippsland for 2010; the expected EBI was calculated for the same land in 2015. The flow of environmental benefits was expected to increase by 13 per cent, from $271 million to $307 million EBI (on the land under contract in the market-based scheme), directly due to the money invested by government in landholders. This was a very useful aspect of the accounts for managers and politicians keen to show the benefits of their decisions.

South East Queensland

The South East Queensland (SEQ) Ecosystem Services Project is a collaborative project between SEQ Catchments, the Queensland Government's Department of Infrastructure and Planning (now the Department of State Development, Infrastructure and Planning) and the University of the Sunshine Coast. The main aims of the project were to develop a framework for assessing the ecosystem services derived from the SEQ region and incorporate this information into NRM policy and planning. Several publications have resulted from the work Maynard, James & Davidson, 2010, 2012; Petter et al., 2013).

The project was funded by the Australian Government's Caring for our Country initiative; the Queensland Departments of Environment and Heritage Protection and State Development, Infrastructure and Planning; the University of Queensland; and the Brisbane city, Moreton Bay and Redland local governments. Technical support was provided by the University of the Sunshine Coast and the Brisbane Regional Environment Council.

The work has focused on identifying, measuring and valuing ecosystem services derived from SEQ. The framework describes four components:

- ecosystem reporting categories
- ecosystem functions
- ecosystem services
- constituents of wellbeing.

A list of the data used in the assessment is available from the project's website;[3] the data layers that were used are also available.[4] The framework recognises that ecosystems contribute to human wellbeing (through the delivery of ecosystem services) and that human wellbeing, which is also derived through social and economic factors, effects the use and condition

---

3   See www.ecosystemservicesseq.com.au/LiteratureRetrieve.aspx?ID=125963
4   See www.ecosystemservicesseq.com.au/LiteratureRetrieve.aspx?ID=120350

of ecosystems. While the SEQ framework does not explicitly use the SEEA or accounting, the aims of the project suggest that accounting could play a role in the future.

Wentworth Group and Condition Accounting for National Resource Management

The Wentworth Group of Concerned Scientists has been advocating the use of environmental accounting since at least 2008 when it published *Accounting for Nature*. Since then, a range of work has been undertaken at both theoretical and practical levels. In the theoretical domain, the Wentworth Group has focused on the development of a metric or 'common currency', which it calls the Environmental Condition Index (Econd).

Econd is not a monetary value; its intent is to allow for the comparison of relative conditions of different environmental assets (e.g. a river with a forest) in different locations, using different metrics. Econd uses the science of reference benchmarking to create an index between zero and 100 that compares the current condition of an asset against a scientific estimate of its natural or potential condition in the absence of significant human alteration.

With funding from a variety of sources, and with support from experts in Australian and state government agencies, the Commonwealth Scientific and Industrial Research Organisation (CSIRO) and the Ian Potter Foundation, the Wentworth Group has undertaken a regional-scale trial of the Accounting for Nature model. The group's ultimate objective is to aggregate this information to create national environmental asset condition accounts. Ten of Australia's 54 regional NRM authorities are involved.

These trials are not yet complete; however, preliminary information is available (Wentworth Group, 2013; Sbrocchi, 2015). The trials have demonstrated that it is practical to apply the concept of asset condition accounting. While the Wentworth Group acknowledges the SEEA, it has not attempted to map its model into the SEEA. In the future, the data that underpin the Econd could be used for accounting for ecosystem extent and condition in the SEEA.

The Great Barrier Reef

The development of land accounts for the Great Barrier Reef region by the ABS led to the production of ecosystem accounts for the region that were published in an information paper (ABS, 2015b). These accounts are consistent with the SEEA-EEA (UN, 2014b). The information paper aimed to test the application of the concepts described in SEEA-EEA, and connect the scientific work undertaken in the region to other environmental and macro-economic indicator accounts compiled by the ABS that build on previous knowledge and data. The ABS accounts, which focus on agriculture, tourism, fishing and aquaculture businesses and their connection to ecosystem services and natural capital, show that it is technically possible to use existing data to create accounts.

Conclusions and Recommendations

Over two decades of work has gone into the development of environmental accounts in Australia. Australia has played a prominent role in the development and testing of the SEEA, which is now recommended for use in Australia (BoM, 2013). However, to date, the systematic and ongoing production of accounts is limited to energy and natural resources accounts (on the national balance sheet) conducted by the ABS, greenhouse gas emissions accounts conducted by the Department of Environment, and water accounts conducted by the ABS and BoM. Despite this, the ongoing development of environmental information systems, combined with increased understandings of the concepts and practices of environmental accounting, means that Australia is well placed to produce accounts when demand grows and the necessary resources are allocated.

Strong partnerships have been a factor in the development of environmental accounting in Australia. These partnerships are required between professions and agencies. Those working together, or at least sharing knowledge and experiences, include the ABS and BoM; the Australian government department responsible for climate change and the environment; the Victorian, Queensland and South Australian state governments; and CSIRO. Various academic institutions and non-government organisations are also involved, such as The Australian National University, University of Queensland, University of Melbourne, University of Sydney and the Wentworth Group of Concerned Scientists.

These partnerships, which encompass a range of professions and specialisations, see geographic information professionals working hand-in-hand with accountants, economists, scientists, statisticians and others.

The production of environmental accounts has improved over time. This is due to greater familiarity with the concept of accounting, as well as refinements in the design of data-compilation activities. Much of the work has involved 'learning by doing', which allows the development of the accounts to be a collaborative process. Repeated production of accounts leads to increased quality. It also allows for efficiencies in compilation processes to be gained through:

- increased knowledge and skills of staff
- ongoing development and use of information technology to support production
- provision of feedback to primary data sources and subsequent improvement in the quality of the primary data
- filling data gaps and deficiencies through the identification or creation of new data sources
- development of useful indicators from the accounts and other data (e.g. gross value of irrigated production per megalitre of water).

The regular production of accounts allows them to be built into policy development, monitoring and evaluation processes.

The value of environmental accounts will be fully realised when they are used in the mainstream decision-making processes of governments, companies and economic activities. Generally, environmental accounts are poorly understood by potential users. Demonstrating how environmental accounts may be applied to policy is a vital first task. In this, the communication of the accounts needs to recognise and target different audiences—policymakers, scientists, economists, accountants, statisticians—and understand the different world views and motivations for accounts. Spatially explicit environmental accounting provides new opportunities for deeper insight and more sophisticated analysis that goes well beyond traditional tabular presentations of accounting data and key summary indicators.

References

Aisbett, E. & Kragt, M. (2010). *Valuing ecosystem services to agricultural production to inform policy design* (Research Report No. 73). Retrieved from crawford.anu.edu.au/research_units/eerh/pdf/EERH_RR73.pdf

ABS (Australian Bureau of Statistics). (2005). *Land management: Eurobodalla shire, NSW, 2003–04* (Catalogue No. 4651.0). Retrieved from www.abs.gov.au/AUSSTATS/abs@.nsf/allprimarymainfeatures/18CAF66FC2574C15CA25710700781E82?opendocument

ABS. (2006). *Land management: Fitzroy and Livingstone shires, Queensland, 2004–05* (Catalogue No. 4651.0). Retrieved from www.abs.gov.au/ausstats/abs@.nsf/mf/4651.0

ABS. (2010). *Land management practices in the Great Barrier Reef catchments, final* (Catalogue No. 4619.0.55.001). Retrieved from www.abs.gov.au/ausstats/abs@.nsf/mf/4619.0.55.001

ABS. (2011). *Land accounts, Great Barrier Reef region, experimental estimates* (Catalogue No. 4609.0.55.001). Retrieved from www.abs.gov.au/AUSSTATS/abs@.nsf/allprimarymainfeatures/6FC78D6DACEF1193CA257D18001339D2?opendocument

ABS. (2012a). *Land accounts, Victoria, experimental estimates* (Catalogue No. 4609.0.55.002). Retrieved from www.abs.gov.au/ausstats/abs@.nsf/mf/4609.0.55.002

ABS. (2012b). *Completing the picture—environmental accounting in practice* (Catalogue No. 4628.0.55.001). Retrieved from www.abs.gov.au/ausstats/abs@.nsf/mf/4628.0.55.001

ABS. (2013). *Land accounts, Queensland, experimental estimates* (Catalogue No. 4609.0.55.003). Retrieved from www.abs.gov.au/ausstats/abs@.nsf/mf/4609.0.55.003

ABS. (2014a). *Land accounts, Great Barrier Reef region, experimental estimates* (Catalogue No. 4609.0.55.001). Retrieved from www.abs.gov.au/ausstats/abs@.nsf/mf/4609.0.55.001

ABS. (2014b). *Water account, Australia* (Catalogue No. 4655.0). Retrieved from www.abs.gov.au/ausstats/abs@.nsf/mf/4610.0

ABS. (2015a). *Australian environmental-economic accounts* (Catalogue No. 4655.0). Retrieved from www.abs.gov.au/ausstats/abs@.nsf/mf/4655.0

ABS. (2015b). *Information paper: An experimental ecosystem account for the Great Barrier Reef region* (Catalogue No. 4680.0.55.001). Retrieved from www.abs.gov.au/ausstats/abs@.nsf/mf/4680.0.55.001

Binning, C., Cork, S., Parry, R. & Shelton, D. (2001). N*atural assets: An inventory of ecosystem goods and services in the Goulburn broken catchment.* Canberra, ACT: CSIRO.

BoM (Bureau of Meteorology). (2013). *Guide to environmental accounting in Australia.* Retrieved from www.bom.gov.au/environment/doc/environmental_accounting_guide.pdf

BoM. (2014). *National water account.* Retrieved from www.bom.gov.au/water/nwa/

Commonwealth of Australia. (2012). *Australia in the Asian century.* Retrieved from www.murdoch.edu.au/ALTC-Fellowship/_document/Resources/australia-in-the-asian-century-white-paper.pdf

Cork, S., Gorrie, G., Ampt, P., Maynard, S., Rowland, P. Oliphant, R., ... Stephens, L. (2012). *Discussion paper on ecosystem services for the Department of Agriculture, Fisheries and Forestry* (Final Report) [Australia21 website]. Retrieved from www.australia21.org.au/wp-content/uploads/2013/08/ALEcosystemservicesR1.pdf

Crossman, N. D., Burkhard, B., Nedkov, S., Willemen, L., Petz, K., Palomo, I., ... Maes, J. (2013). A blueprint for mapping and modelling ecosystem services. *Ecosystem Services 4,* 4–14, doi.org/10.1016/j.ecoser.2013.02.001

DEE (Department of Environment and Energy). (n.d.-a). *National plan for environmental information initiative.* Retrieved from www.environment.gov.au/science/national-plan-environmental-information

DEE. (n.d.-b). *National greenhouse gas and energy reporting.* Retrieved from www.environment.gov.au/climate-change/greenhouse-gas-measurement/nger

DEE. (n.d.-c). *National greenhouse and energy reporting publications.* Retrieved from www.environment.gov.au/climate-change/greenhouse-gas-measurement/publications

Department of Environment, Water, Heritage and Arts. (2009). *Ecosystem services: Key concepts and applications.* Retrieved from www. environment.gov.au/biodiversity/publications/ecosystem-services-key-concepts-and-applications

Eigenraam, M., Chua, J. & Hasker, J. (2013). *Environmental-economic accounting: Victorian experimental ecosystem accounts* (Version 1.0) [Department of Sustainability and Environment website]. Retrieved from ensym.dse.vic.gov.au/docs/Victorian%20 Experimental%20Ecosystem%20Accounts,%20March%202013.docx

European Commission. (2014). *SEEA extensions and applications.* Retrieved from unstats.un.org/unsd/envaccounting/ae_white_cover.pdf

Expert Meeting on Ecosystem Accounts. (2012, 16–18 May). *List of documents.* Retrieved from unstats.un.org/unsd/envaccounting/seea LES/egm2/lod.htm

Gillespie, R., Dumsday, R. & Bennett, J. (2008). *Estimating the value of environmental services provided by Australian farmers.* Surrey Hills, NSW: Australian Farm Institute.

Gleeson-White, J. (2014). *Six capitals.* Sydney, NSW: Allen & Unwin.

Godfrey, J. M. & Chalmers, K. (2012). *Water accounting: International approaches to policy and decision-making.* Cheltenham, UK: Edward Elgar. doi.org/10.4337/9781849807500

Hawke, A. (2009). *The Australian Environment Act—report of the independent review of the Environment Protection and Biodiversity Conservation Act 1999* [Department of the Environment, Water, Heritage and the Arts website]. Retrieved from www.environment.gov. au/system/files/resources/5f3fdad6-30ba-48f7-ab17-c99e8bcc8d78/ files/final-report.pdf

Maynard, S., James, D. & Davidson, A. (2010). The development of an ecosystem services framework for south east Queensland. *Environmental Management 45*(5), 881–95. Retrieved from link. springer.com/article/10.1007%2Fs00267-010-9428-z

Maynard, S., James, D. & Davidson, A. (2012). An adaptive participatory approach for developing an ecosystem services framework for south east Queensland, Australia. *International Journal of Biodiversity Science, Ecosystem Services and Management 7*(3), 182–89. doi.org/10.1080/21 513732.2011.652176

Morton, S. & Tinney, T. (2011). *Independent review of Australian Government environmental information activity.* Retrieved from www. environment.gov.au/system/files/pages/8d3f2610-c336-4e47-aaa7-f3d2b879b905/files/eia-review-discussion-paper.pdf

Obst, C. & Vardon, M. (2014). Recording environmental assets in the national accounts. *Oxford Review of Economic Policy 30*, 124–44. doi.org/10.1093/oxrep/gru003

Petter, M., Mooney, S., Maynard, S., Davidson, A., Cox, M. & Horosak, I. (2013). A methodology to map ecosystem functions to support ecosystem services assessments. *Ecology and Society 18*(1), 31. Retrieved from www.ecologyandsociety.org/vol18/iss1/art31/

Pittock, J., Cork, S. & Maynard, S. (2012). The state of the application of ecosystems services in Australia. *Ecosystem Services 1*, 111–20. doi.org/10.1016/j.ecoser.2012.07.010

Russell-Smith, J., Ada, J., Barker, P., Campbell, A., Cork, S., Costanza, B., … Yates, C. (2013). *Ecosystem services and livelihood opportunities for Indigenous rural communities in savanna landscapes.* Retrieved from www.aceas.org.au/Ecosystem_services_and_Indigenous_livelihoods-Report.pdf

Sbrocchi, C. (2015). *Australian regional environmental accounts trial: Report to NRM regions Australia.* Retrieved from wentworthgroup. org/2015/03/report-to-nrm-regions-australia/2015/

Smith, R. (2014). *Users and uses of environmental accounts: A review of selected developed countries.* Washington, DC: World Bank. Retrieved from www.wavespartnership.org/sites/waves/files/documents/PTEC1-%20 Users%20and%20Uses%20of%20Environmental%20Accounts.pdf

Stoeckl, N., Hicks, C. C., Mills, M., Fabricius, K., Esparon, M., Kroon, F., … Costanza, R. (2011). The economic value of ecosystem services in the Great Barrier Reef: Our state of knowledge. *Annals of the New York Academy of Sciences 1219*, 113–33. doi.org/10.1111/j.1749-6632.2010.05892.x

Stoneham, G., O'Keefe, A., Eigenraam, M. & Bain, D. (2012). Creating physical environmental asset accounts from markets for ecosystem conservation. *Ecological Economics 82*, 114–22. doi.org/10.1016/j.ecolecon.2012.06.017

Straton, A. & Zander, K. (2009). *The value of Australia's tropical river ecosystem service*s. Land and Water Australia, Tropical Rivers and Coastal Knowledge, Charles Darwin University and CSIRO, Darwin.

Thackway, R. & Lesslie, R. (2006). Reporting vegetation condition using the vegetation assets, states and transitions (VAST) framework. *Ecological Management and Restoration 7*, S53–S62. doi.org/10.1111/j.1442-8903.2006.00292.x

Tovey, J. P. (2008). Whose rights and who's right? Valuing ecosystem services in Victoria, Australia. *Landscape Research 33*, 197–209. doi.org/10.1080/01426390801908426

UN (United Nations). (2012). *System of environmental-economic accounting for water*. Retrieved from unstats.un.org/unsd/envaccounting/seeaw/seeawaterwebversion.pdf

UN. (2014a). *System of environmental-economic accounting—central framework*. Retrieved from unstats.un.org/unsd/envaccounting/seeaRev/SEEA_CF_Final_en.pdf

UN. (2014b). *System of environmental-economic accounting experimental ecosystem accounting*. Retrieved from unstats.un.org/unsd/envaccounting/seeaRev/eea_final_en.pdf

van Dijk, A., Mount, R., Gibbons, P., Vardon, M. & Canadell, P. (2014). Environmental reporting and accounting in Australia: Progress, prospects and research priorities. *Science of the Total Environment 473–74*, 338–49 dx.doi.org/10.1016/j.scitotenv.2013.12.053

Vardon, M. (2012). The system of environmental-economic accounting for water: Development, implementation and use. In J. M. Godfrey & K. Chalmers (Eds.), *Water accounting: International approaches to policy and decision-making* (pp. 32–57). Cheltenham, UK: Edward Elgar.

Vardon, M., Burnett, P. & Dovers, S. (2016). The accounting push and the policy pull: Balancing environment and economic decisions. *Ecological Economics 124*, 145–52. doi.org/10.1016/j.ecolecon.2016.01.021

Vardon, M., Lenzen M., Peevor, S. & Creaser M. (2007). Water accounting in Australia. *Ecological Economics 61*(4), 650–59. doi.org/10.1016/j. ecolecon.2006.07.033

Wentworth Group of Concerned Scientists. (2008). *Accounting for nature*. Retrieved from wentworthgroup.org/2008/05/accounting-for-nature-a-model-for-building-the-national-environmental-accounts-of-australia/2008/

Wentworth Group of Concerned Scientists. (2013). *Initial observations on the Australian proof of concept regional environmental asset condition trials.* Retrieved from wentworthgroup.org/2013/08/initial-observations-of-regional-environmental-accounts-proof-of-concept-trial/

15

Elephants in the Kitchen: Responding to the Challenge of Rapidly Changing Climate and Land Use

Brendan Mackey

Key Points

- The impacts of human-forced climate change will continue to be felt for millennia, irrespective of our success or failure to mitigate greenhouse gas emissions.
- Therefore, climate change adaptation must be understood as a 'forever' activity and mainstreamed in policy, planning and decision-making.
- In parallel, we are witnessing unprecedented land use change in terms of extent and intensification that is transforming the land surface, together with subsurface processes, in ways as profound as climate change.
- Policies and programs that aim to reduce emissions of greenhouse gases from the land sector will continue to be a key component of Australia's response to its Paris Agreement commitments.
- We have the data and information to address these emerging pressures; however, we lack the public understanding, political will and national policy to require their use.

- State and territory governments must take the necessary step change. There is a need for national leadership by the Australian Government if the new generation of data, knowledge products and decision support tools are to be developed, and the policies that drive their application in support of agreed national land use goals put in place.

Introduction

An elephant in the kitchen, or any other room for that matter, is an idiom for an obvious problem that no one wants to discuss. Here, I discuss some elephants that public policy currently ignores: the synergistic effects on the land sector of climate change and land use change, and the implications for ecologically sustainable development. I then consider the kind of data, information products, decision support tools and policy responses that are needed if we are to have the informed land use planning and management that these problem warrant.

The extent to which climate change presents profound risks to the economy has been on the record at least since the work of William Cline (1992) and, more popularly, since the Stern (2006) review, which estimated that without mitigation, the costs would be equivalent to losing 5–20 per cent of global GDP each year (i.e. US$3.9–$15.6 trillion in 2014), now and forever, versus around 1 per cent of global GDP each year to avoid harm (i.e. US$0.8 trillion in 2014) (World Bank, 2015). From the public sector perspective, more recent estimates suggest that 6 degrees Celsius of warming represent present-value losses worth US$43 trillion, which is about 30 per cent of the entire stock of the world's manageable assets (Economist Intelligence Unit, 2015). These economic statistics point to the magnitude of the problems we face in all sectors, including land sectors, in mitigating emissions and adapting to a rapidly changing climate.

The land sector is multifaceted when it comes to climate change. It functions as both a source of emissions and a sink; terrestrial ecosystems naturally exchange carbon dioxide with the atmosphere and intensive land use depletes organic carbon stocks. About 27 per cent of the total accumulated anthropogenic greenhouse gas emissions are from the land sector (compared with 35 per cent from coal); they constitute about 10 per cent of current annual global emissions, and there is still about four times the atmospheric stock of carbon remaining in terrestrial ecosystems

(Global Carbon Project, 2014; Mackey et al., 2013). However, the land is also the place where people live, harvest fresh water and grow most (90 per cent) of their calorie intake (with aquaculture, freshwater and marine fisheries supplying about 10 per cent) (Nellemann et al., 2009). Climate change also affects the functioning of natural, semi-natural and agro-industrial systems, with implications for species, including invasive species ecosystems and agricultural productivity (Intergovernmental Panel on Climate Change [IPCC], 2014).

Climate Change is 'Forever'

The idea of the effects of climate change being 'forever', as implied by the Stern review, is scientifically valid. Due to the extraordinarily long atmospheric lifetime of a pulse of carbon dioxide, together with lag effects in the Earth system (especially the oceans), the impacts of human-forced climate change will continue to be felt for millennia, irrespective of our success or failure to mitigate greenhouse gas emissions (Mackey et al., 2013). However, this is not to say that we should abandon efforts to mitigate greenhouse gas emissions. Deep cuts in emissions are required to meet the Paris Agreement (United Nations, 2015) commitment to holding the increase in the global average temperature to well below 2 degrees Celsius above pre-industrial levels, and pursuing efforts to limit the temperature increase to 1.5 degrees Celsius above pre-industrial levels. Conversely, a business-as-usual approach could lead to increases of more than 6 degrees Celsius as we enter the twenty-second century (Brown & Caldeira, 2017). Notwithstanding our success or failure to mitigate, climate change adaptation must be understood as a 'forever' activity and mainstreamed into policy, planning and decision-making at all levels of governance and in all sectors. The prospect of continuously changing climate impacts also has significant implications for data management, as it implies ongoing monitoring and adaptive management responses in light of new information.

Land Use Change is Unprecedented

In parallel with rapid human-forced climate change, we are witnessing unprecedented land use change in terms of the rapid extension of modern economic activities onto previously natural lands, and the intensification of land use through agro-industrialisation (Goldewijk, Beusen, Van Drecht & De Vos, 2011). This rapid land use change is being catalysed by a combination of financial, globalisation and technological innovations, as we witness a national increase in agro-industrial and mining enterprises of increasing magnitude. Examples include coal seam gas mining (Australian Broadcasting Commission, 2012), coalmining (Department of State Development, 2015) and irrigated farming schemes (Department of Agriculture and Fisheries, 2015). Further indications of the land use changes that lie ahead are articulated in the White Paper on Developing Northern Australia, and the new National Water Infrastructure Development Fund that, in addition to accelerating investment in water infrastructure, has more than $5 billion available for other forms of infrastructure—much of which will directly or indirectly support land use change in the region (Australian Government, 2015).

The Low Carbon Economy

To stabilise atmospheric concentrations of greenhouse gases at a level that gives a greater than 50 per cent chance of limiting warming to below 2 degrees Celsius above pre-industrial levels, only about an additional 1,000 gigatons of carbon dioxide (CO_2) can be emitted; this amounts to about 30 years' worth of current annual global emissions (IPCC, 2013). Given the deep cuts needed in emissions to meet this target, it is not surprising that a range of mitigation policy approaches are being explored, including market-based mechanisms. Carbon pricing mechanisms currently exist in around 40 countries, including emission trading schemes, carbon taxes, payments for emission reductions and the purchase of offsets (Economist Intelligence Unit, 2015; World Bank, 2016). While the concept of a comprehensive global carbon market is appealing in theory, it is unlikely in practice. The most likely scenario is that the 'bottom-up' approach underpinning the Paris Agreement—whereby each country makes nationally determined emission reduction commitments and employs mitigation approaches and policies that are tailored to national circumstances—will continue. The result, for better or worse, will be a diverse policy landscape in which some jurisdictions have a carbon price on certain emissions, but not others; a cohort of jurisdictions will have no carbon price on any emissions, but may have some international linking of carbon pricing mechanisms and continued trading of offsets consistent with the provisions of the Paris Agreement.

Given the Paris Agreement commitment to holding the increase in the global average temperature to well below 2 degrees Celsius above pre-industrial levels, and pursuing efforts to limit the temperature increase to 1.5 degrees Celsius above pre-industrial levels, the pressure is now on to reduce emissions from all sources, including the land sector, which is especially significant for Australia's national greenhouse gas accounts. Australia's 2012 fossil fuel emissions were 31 per cent above its 1990 levels; however, when land use, land use change and forestry are included, total emissions had only increased by 2.4 per cent. The difference arises because emissions can be deducted, in accounting terms, by 'withdrawals' from the atmosphere into ecosystems; land sector credits offset 28.6 per cent of Australia's industrial emissions over this period. Given the significance of the land sector to Australia meeting mitigation reduction targets, we can anticipate that the land sector will face ongoing and new pressures generated from Australia's Paris Agreement commitments.

In addition to carbon credit–generating activities, such as afforestation, reforestation and avoided emissions through conservation, new agricultural enterprises are already emerging to service the low carbon economy. Biofuel is now mandated at some level in around 64 countries, including the European Union (EU) (which has mandated 5–7.5 per cent renewable content by 2020), 13 countries in the Americas, 12 in the Asia–Pacific region, 11 in Africa and the Indian Ocean and two from non-EU countries in Europe. Global demand will be affected by targets set in the United States, China and Brazil; these countries have set targets in the 15–25 per cent range by 2020–22 (Biofuels Digest, 2014). Another emerging, and potentially disruptive, industry is biomaterials made from bio-based or bio-derived polymers from natural renewable resources, such as corn, soy, potatoes and sugar cane, rather than petroleum feedstocks (PolyOne, 2008).

Demands for new land uses to service the low carbon economy will be driven by a combination of public policy, such as national mitigation reduction targets, and private investments. These demands will compete with existing land uses, especially food production, for fertile land, water and capital. The synergistic effects of a rapidly changing climate and new land uses promise to be highly economically, ecologically and socially disruptive, with prospects for surprises and unplanned feedback in what is, from a scientific perspective, a non-linear and complex socio-ecological system.

Data, Information and Decision Support

The kinds of land use problems we face are becoming increasingly complicated due to the synergistic interactions of climate change, technological innovation, influx of foreign capital (Sanyal, 2014) and new land-based economic activities, such as coal seam gas and biofuels. Fortunately, at all three levels of government, Australia has made great advances in land-related data and its online availability, including data on surface and groundwater resources, natural hazards, remotely sensed land cover and high-resolution continental soil grids (e.g. Terrestrial Ecosystem Research Network, 2015). We also have ready access to the world's most advanced current and future climate data, including regionally scaled projections (Commonwealth Scientific and Industrial Research Organisation, 2015; Department of Premier and Cabinet, 2015).

However, there remains a gap in processing many of the available data into information products that are useful for policy and decision-making, such as catchment condition assessments (e.g. Worboys, Good & Spate, 2011). Further, decision-making in a climate-changed world requires more sophisticated approaches, such as the use of scenario modelling to help identify policy options that are robust under plausible alternative futures, which can account for accumulated and long-term impacts on our scarce water and fertile soil resources, among other things.

However, we face a more fundamental problem than lack of the right information and decision support tools. Land development policies and proposals continue to be rolled out without proper reference to the available data and information, and large-scale industrial developments are being approved where the data and system understanding needed for sustainability planning are known to be lacking. The states and territories have conventionally been primarily responsible for planning and land management. It follows that current failures should be laid primarily at their feet, rather than with the Australian Government. Nonetheless, the Australian Government has external affairs powers that enable it to give effect to international treaty obligations through national policy, along with powers over interstate trade, commerce and corporations. To give effect to Australia's obligations under relevant international conventions, including the UNFCCC, the Convention on Biological Diversity and the Convention on Wetlands (known as the Ramsar Convention), measures need to be taken concerning land use management.

The Australian Government has an acknowledged role in helping to coordinate and harmonise state and territory government policies regarding matters that serve the national good. These arrangements are often negotiated at the Council of Australian Governments and facilitated through non-coercive means such as grants. The water resources sector shows how national objectives can be advanced on natural resource management issues; although some problems remain, the states reached an agreement with the Australian Government. The Australian Government used its constitutional powers, including external affairs, to enact the *Water Act 2007*. This Act established the Murray–Darling Basin Authority, which has the functions and powers, including enforcement powers, needed to help ensure that Basin water resources are managed in an integrated and sustainable way. Given the pressures on the Australian landscapes from the imperative to mitigate greenhouse gas emissions, disruptive influences of emerging land use activities and impacts of a rapidly changing climate, among other things, such cooperative arrangements will be needed in the coming years.

Recommendations

We have unprecedented data, information and decision support tools available for improved land use decision-making at all levels of government. Information communication technology (ICT) is making astonishing advances with high-performance computers, mass-data storage and internet connectivity approaching transformational thresholds. Now is the time to consolidate our excellent progress, take advantage of emerging ICT capacities and generate the kinds of advanced information and decision support needed in the coming decades for ecologically sustainable land use policy, planning and management.

However, in the absence of both a broader base of recognition of the seriousness and urgency of these problems and their significance for long-term ecological sustainability and a corresponding surge in the political will, it is unlikely that the investment of funds needed to drive this step change in capacity will be forthcoming. While state and territory governments must take on the required step change themselves, there can be no doubt that the Australian Government has a leadership role to play. This role could be realised through the collaborative formulation of a national land use strategy that drives the development of the required data, information products and decision support tools, and requires their application at all levels of government to meet agreed land use goals.

References

Australian Broadcasting Commission. (2012, 28 June). *The coal seam gas rush. ABC News Online.* Retrieved from www.abc.net.au/news/specials/coal-seam-gas-by-the-numbers/

Australian Government. (2015). *Our north, our future: White paper on developing northern Australia.* Canberra, ACT: Australian Government. Retrieved from northernaustralia.gov.au/files/files/NAWP-FullReport.pdf

Biofuels Digest. (2014, December 31). Biofuels mandates around the world: 2015. *Biofuels Digest.* Retrieved from www.biofuelsdigest.com/bdigest/2014/12/31/biofuels-mandates-around-the-world-2015/

Brown, P. T. & Caldeira, K. (2017). Greater future global warming inferred from Earth's recent energy budget. *Nature, 552,* 45–50. doi.org/10.1038/nature24672

Cline, W. R. (1992). *The economics of global warming.* Washington, DC: Institute for International Economics.

Commonwealth Scientific and Industrial Research Organisation. (2015). *Climate change in Australia.* Canberra, ACT: Department of Environment and Bureau of Meteorology. Retrieved from www.climatechangeinaustralia.gov.au/en/

Department of Agriculture and Fisheries. (2015). *Flinders Gilbert agricultural zone.* Retrieved from www.daf.qld.gov.au/business-trade/development/industry-development/flinders-gilbert-agricultural-zone

Department of Premier and Cabinet. (2015). *Climate futures for Tasmania.* Retrieved from acecrc.org.au/climate-futures-for-tasmania/

Department of State Development. (2015). *Carmichael coal mine and rail project—project overview.* Retrieved from www.statedevelopment.qld.gov.au/assessments-and-approvals/carmichael-coal-mine-and-rail-project.html

Economist Intelligence Unit. (2015). *The cost of inaction: Recognising the value at risk from climate change*. Retrieved from www.eiuperspectives.economist.com/sites/default/files/The%20cost%20of%20inaction_0.pdf

Global Carbon Project. (2014). *Global carbon budget 2014—an annual update of the global carbon budget and trends*. Retrieved from www.globalcarbonproject.org/carbonbudget/index.htm

Goldewijk, K. K, Beusen, A., Van Drecht, G. & De Vos, M. (2011). The HYDE 3.1 spatially explicit database of human-induced global land-use change over the past 12,000 years. *Global Ecology and Biogeography, 20,* 73–86. doi.org/10.1111/j.1466-8238.2010.00587.x

IPCC (Intergovernmental Panel on Climate Change). (2013). Summary for policymakers. In T. F. Stocker, D. Qin, G.-K. Plattner, M. Tignor, S. K. Allen, J. Boschung, A. Nauels, Y. Xia, V. Bex & P. M. Midgley (Eds.), *Climate change 2013: The physical science basis* (Section E.8 p. 27). Cambridge, UK: Cambridge University Press.

IPCC. (2014). Summary for policymakers. In C. B. Field, V. R. Barros, D. J. Dokken, K. J. Mach, M. D. Mastrandrea, T .E. Bilir, … L.L. White (Eds.), *Climate change 2014: Impacts, adaptation, and vulnerability* (Part A: Global and sectoral aspects) (pp. 1–32). Cambridge, UK: Cambridge University Press.

Mackey, B., Prentice, I. C., Steffen, W., House, J. I., Lindenmayer, D., Keith, H. & Berry, S. (2013). Untangling the confusion around land carbon science and climate change mitigation policy. *Nature Climate Change, 3,* 552–57. doi.org/10.1038/nclimate1804

Nellemann, C., MacDevette, M., Manders, T., Eickhout, B., Svihus, B., Prins, A. G. & Kaltenborn, B. P. (Eds.). (2009). *The environmental food crisis: The environment's role in averting future food crises* [GRID-Arendal website]. Retrieved from www.grida.no/publications/154

PolyOne. (2008). *Biomaterials development in the polymer industry* (Technical Bulletin). Retrieved from www.polyone.com/files/resources/Biomaterials_Development_in_the_Polymer_Industry.pdf

Sanyal, K. (2014). *Foreign investment in Australian agriculture* (Research Paper Series 2013–14). Canberra, ACT: Department of Parliamentary Services.

Stern, N. (2006). *Stern review on the economics of climate change*. London, UK: HM Treasury.

Terrestrial Ecosystem Research Network. (2015). *Terrestrial Ecosystem Research Network*. Retrieved from www.tern.org.au/

United Nations. (2015). *The Paris agreement*. Retrieved from unfccc. int/files/essential_background/convention/application/pdf/english_paris_agreement.pdf

Worboys, G. L., Good, R. B. & Spate, A. P. (2011). *Caring for our Australian Alps catchments* (Technical Report). Canberra, ACT: Department of Climate Change and Energy Efficiency.

World Bank. (2015). *Gross domestic product 2014*. World Development Indicators database, World Bank. Retrieved from databank.worldbank. org/data/reports.aspx?source=world-development-indicators

World Bank. (2016). *Pricing carbon*. Retrieved from www.worldbank.org/en/programs/pricing-carbon

Part 5 – Visions for the Future

16

The Future for Land Use Mapping: National E-Infrastructure, Modelling Analytics, Synthesis and Securing Institutional Capacity

Tim Clancy, Brett A. Bryan and Siddeswara M. Guru

Key Points

- Land use and land use change are central to our understanding of human impacts on the environment; they underpin policy and programs to support productivity improvements, protection and land development.
- Increased demands for information and improved technologies provide a driver to improve our ability to characterise, track and model future trends in land use.
- Improved technologies, applications and synthetic approaches, such as increased analytical power through harnessing cloud computing and high-performance processing, provide significant opportunities.
- Some of these recent technologies, both within and outside the land use mapping context, are presented to map a vision for the future.

- The establishment of a land use and land resources centre—a virtual organisation operating on a collaborative basis and drawing on the expertise of Australian agencies and research institutions with allied interests and objectives—would be an excellent way of securing the institutional capabilities to deliver on this vision.

Introduction

Land use and land use change are central to our understanding of human impacts on the environment (e.g. Foley et al., 2005); they underpin policy and programs to support productivity improvements, protection and use of our natural resources and land development (Clancy & Lesslie, 2013). Increased demands for information and improved technology provide both a driver of change to existing approaches and an opportunity to improve our ability to characterise, track and model future trends in land use (Bryan, 2013). In this chapter, we explore some of these opportunities, focusing on current advances in e-infrastructure and examples of their application. A goal of this paper is to look at ways to address one of Rob Lesslie's key concerns: the ad hoc nature of both the funding and delivery of crucial national and regional land use products. Some of these recent technologies, both within and outside the land use mapping context, are presented to map a vision for the future, including the establishment of a centre for land use and land resources.

Current and Future Drivers

Land use decisions have a direct impact on our food production systems, natural environment and communities, and are central to many current debates (Lesslie & Mewett, 2013). It is relatively easy to identify a broad range of drivers, both internationally and domestically, for access to high-quality land use information. High quality is defined as:

- accurate representation of on-ground land use
- precise measures of trends in land use change
- sound future scenarios
- management-appropriate scales.

Issues such as food security, agriculture productivity improvements, (ecologically) sustainable development, biosecurity and disaster-degradation mitigation are dependent on understanding land use. These issues provide important contextual information for the scientific and policy analyses undertaken by international bodies, such as the United Nations Framework Convention on Climate Change, Convention on Biological Diversity and United Nations Convention to Combat Desertification. At the national level, elements of these international drivers are relevant; so too are specific priorities, such as current aspirations to foster development of northern Australia, land-based carbon capture programs and the management of nutrient run-off onto the Great Barrier Reef.

This diversity of drivers is both a strength and a weakness in the development of robust, consistent and high-quality land use information. The strength is in the value placed on the information provided; the weakness is that there is no single issue that drives investment, nor a single agency or single level of government with ultimate responsibility.

Increasing Computational Demands from Land Science

The drivers of and solutions to climate change, food and energy security, natural resource management and biodiversity conservation all reside within extremely complex socio-ecological systems that demand the integrated assessment and modelling of multidisciplinary 'big' data (Bryan, 2013).

Increasingly complex and interconnected problems confronting land use scientists, managers and policymakers in the land system require new technological approaches. The solution to complex global challenges requires information on the management of land use at high spatial and temporal resolutions over continental or global extents. However, until recently, computing capacity was a barrier to the sort of large-scale, high-resolution modelling required (Zhao et al., 2015). Bryan, Crossman, King and Meyer (2011) provided examples of the type of complex analyses required in dealing with issues of sustainable agriculture. Often, we are interested in the analysis of potential future landscapes, which require multiple objectives, multiple scenarios, complex prioritisation and quantification of a range of impacts. Even when dealing with constrained geographic areas, such analyses require large amounts of input data and significant computing power.

High-Performance Computing

The computational demands of integrated modelling across space and time (hindcasting and forecasting), as well as the realities of the complex social-ecological systems required to address global environmental challenges, are unlikely to be met by traditional geographic information system (GIS) tools that are largely constrained to the desktop or local server. Bryan (2013) evaluated the potential of a range of high-performance computing (HPC) hardware and software tools to overcome these computational barriers and found clear potential for spectacular gains in performance. He conducted four realistic simulation experiments using:

a. Arc Macro Language (AML) GIS script on a single central processing unit (CPU)

b. Python/NumPy on 1–256 CPU cores

c. Python/NumPy on 1–64 graphics processing units (GPUs) with high-level PyCUDA abstraction (GPUArray)

d. Python/NumPy on 1–64 GPUs with low-level PyCUDA abstraction (ElementwiseKernel).

The gains in performance from the GIS implementation, which effectively took 15.5 weeks to run, were marked, with speed gains of 59× for scenario b) compared to a). More impressively, there was a 4,881× increase using ElementwiseKernel c) compared to a) with, at best, the ElementwiseKernel module in parallel over 64 GPUs achieving a speed-up of >60,000× d) compared to a).

Open source tools, such as Python, applied across a spectrum of HPC resources, offer transformational and accessible performance improvements for integrated assessment and modelling. By reducing the computational barrier, HPC can lead to a step change in modelling sophistication, including the better representation of uncertainty and, perhaps, new modelling paradigms (Bryan, 2013).

There are currently major costs with migration to new hardware and software environments in the HPC environment; however, as Bryan (2013) pointed out, if researchers can be freed of the computational constraints with access to HPC, they can develop new approaches to addressing global environmental challenges, which augurs well for an exciting future.

Use of a Scientific Workflow Approach

A driver of e-infrastructure development is the move from data storage and computer-focused analyses (generally limited by local infrastructure) to a focus on methodologies as tools and data, where these are brought together and cloud computing and storage systems are fully harnessed (Francis, 2015). Scientific workflow technology has become popular by providing a high-level environment that can automate, manage and execute various steps in scientific research, while also having the ability to store and track provenance information. Scientific workflows provide a powerful unifying platform that allows scientists to arbitrarily build complicated applications by combining predefined components that may be implemented in different programming languages (Ludäscher et al., 2006). Once the workflow is built, it can be re-used and re-executed with minimal effort. These intrinsic capabilities of a workflow system with provenance tracking functionality would improve the reproducibility of experiments and encourage the sharing of experimental processes and results. Workflow systems offer a broad range of components that perform tasks, ranging from acquiring data from sensors, querying databases, data-mining and visualisation through to execution of arbitrary applications.

The Collaborative Environment for Ecosystem Science Research and Analysis (CoESRA) system developed by the Terrestrial Ecosystem Research Network is a web-enabled, virtual desktop environment, running on a cloud infrastructure. This is an example of where a workflow approach (effectively an analysis tool) can be integrated into national computer and storage infrastructure that addresses the barriers to entry to HPC and also provides a scalable and flexible virtual laboratory environment.[1] Users can access a virtual desktop environment to build, execute and share

---

1   This system is available online from www.coesra.org.au

workflow-based scientific analyses and syntheses activities. The system supports Australian Access Federation as a login mechanism (Guru et al., 2015).

The CoESRA platform is built on Queensland Research and Innovation Services Cloud (QRISCloud) infrastructure, which enables access to data sources that are available from web services, large storage, distributed computing infrastructure and analysis tools to implement cases from the ecosystem science community. Using CoESRA, scientists can make their analyses repeatable and sharable, which may improve the uptake of scientific outcomes (Guru et al., 2015).

Several innovative concepts were developed as part of the system to reduce the barriers to using cloud and distributed computing infrastructure to run complex, data-intensive analyses. The system has provided new methods for supporting research, including the ability to provide computing infrastructure on a virtual desktop via a web browser, run sophisticated analysis software (e.g. Marxan in Kepler workflow) and develop shareable workflows to support an ecosystem research and investigation framework. The ability to create reusable experiments as a set of interconnected tasks, and the ability to submit or invoke jobs on HPC, will make the concept of workflow very useful to promoting open science and bringing transparency to scientific experiments.

Land Use and Land Resources Centre

The increased analytical power available through the harnessing of cloud computing and high-performance processing (as well as enhanced models that blend data from multiple sources and work across spatial and temporal scales) provides an opportunity to take a more strategic approach to land use mapping and change monitoring. Coupled with the use of appropriate workflow approaches that enhance the ability to share, collaborate, automate, repeat and repurpose land use mapping systems, these advances present an opportunity to address the ongoing need for high-quality, scale-appropriate land use information products that are cost efficient and have high levels of precision, accuracy and currency.

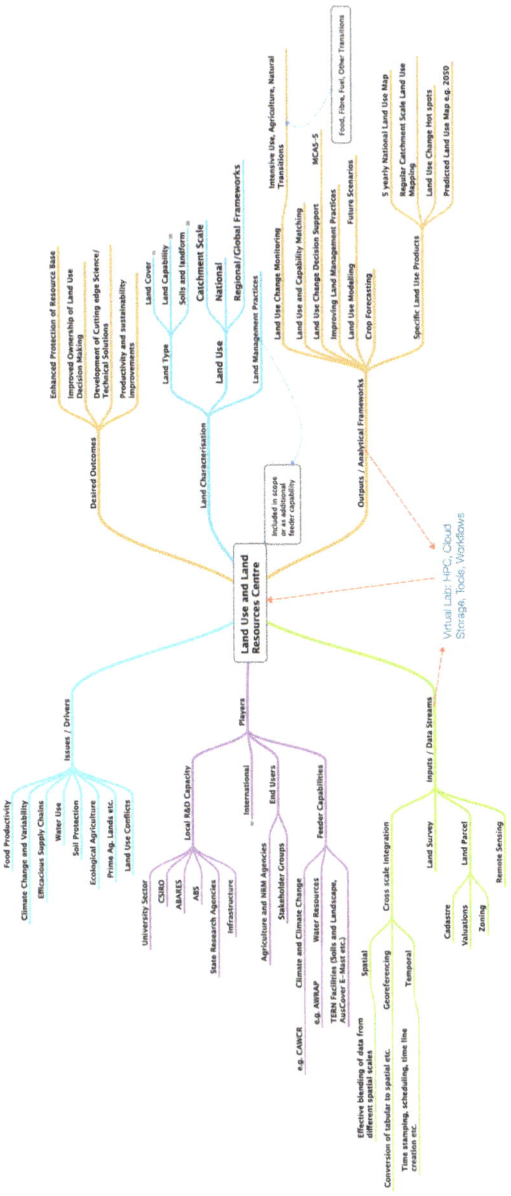

Figure 16.1: Organising framework for a land use and land resources centre designed to bring different capabilities together and leverage off existing investments by various stakeholders.

Note: This is a conceptual diagram. Moving clockwise from the top left: summary of 1) key drivers; 2) outcomes sought in policy and programs; 3) types of land resource layers required for analyses; 4) outputs the centre could produce; 5) required diversity of data to be accessed; and 6) stakeholders and collaborators needed. Source: Adapted from Clancy & Lesslie (2013).

The concept of a land use and land resources centre was scoped in recent times from a food security perspective (Clancy & Lesslie, 2013). Beyond the central issue of food security, it was found that a broad range of needs could be addressed more directly and effectively by focusing specifically on improving Australia's capacity to analyse and track land use change, particularly in relation to our productive land assets. This would address stakeholder groups' demands, such as calls from the agricultural sector for improved data and analysis to assess land use change and agricultural production potential. A land use and land resources centre would provide an organising framework for bringing different capabilities together, while leveraging off existing investments (Figure 16.1).

The proposed land use and land resources centre would be a virtual organisation, operating on a collaborative basis and drawing on the expertise of Australian agencies and research institutions with allied interests and objectives. It would conduct an Australian program of work linked to international efforts on agricultural land use and food security analysis. In doing so, it could make an important contribution to global food security in terms of land use and land resources risk assessments, particularly the development of protocols for tracking and forecasting change, and the development of tools to assist decision-making.

Recommendations

1. The existing frameworks for national and state collaboration should be maintained and strengthened, including the direct involvement of science agencies such as the Commonwealth Scientific and Industrial Research Organisation (CSIRO), universities and the National Committee on Land Use and Management Information (NCLUMI).

2. It should be recognised that opportunities for harnessing HPC needs are central to discussions of national land use programs, including the NCLUMI.

3. It is necessary to move to a strategic approach that involves regular national and regional updates of land use mapping based on a consistent funding stream that recognises the broad range of uses and needs for land use products.

4. The concept of a land use and land resources centre should be further explored to address recommendations 1–3.

5. If such a physical and virtual institution were to be established, naming it the (Rob) Lesslie Land Use and Land Resource Centre would be a fitting memorial to Rob's important work in this area.

References

Bryan, B. A. (2013). High-performance computing tools for the integrated assessment and modelling of social–ecological systems. *Environmental Modelling and Software 39*, 295–303. doi.org/10.1016/j.envsoft. 2012.02.006

Bryan, B. A., Crossman, N. D., King, D. & Meyer, W. S. (2011). Landscape futures analysis: Assessing the impacts of environmental targets under alternative spatial policy options and future scenarios. *Environmental Modelling and Software 26*, 83–91. doi.org/10.1016/j. envsoft.2010.03.034

Clancy, T. F. & Lesslie, R. G. (2013). *A scoping assessment for a national research centre addressing land use and food security issues.* Canberra, ACT: Department of Agriculture, Fisheries and Forestry.

Foley, J. A., DeFries, R., Asner, G. P., Barford, C., Bonan, G., Carpenter, S. R., … Snyder, P. K. (2005). Global consequences of land use. *Science 309*(5734), 570–74. doi.org/10.1126/science.1111772

Francis, R. (2015, July). *Research methods—the great integrator.* Paper presented at the Australian eResearch Organisation's 6th National Forum: Roadmapping Australia's eResearch Future, Canberra. Retrieved from aero.edu.au/wp-content/uploads/2015/08/AeRO_ Forum_2015_Rhys_Francis_presentation.pdf

Guru, S. M., Nguyen, H. A., Banihit, S., Mulholland, M., Olsson, K. & Clancy, T. (2015). *Development of cloud-based virtual desktop environment for synthesis and analysis for ecosystem science community.* International Workshop on Science Gateways, eResearch Australasia. Retrieved from eresearchau.files.wordpress.com/2015/11/guru_coesra _iwsg.pdf

Lesslie, R. & Mewett, J. (2013). *Land use and management: The Australian context* (Research Report 13.1). Canberra, ACT: Australian Bureau of Agricultural and Resource Economics and Sciences.

Ludäscher, B., Altintas, I., Berkley, C., Higgins, D., Jaeger, E., Jones, M., … Zhao, Y. (2006). Scientific workflow management and the Kepler system. *Concurrency and Computation: Practice and Experience 18*, 1039–65. doi.org/10.1002/cpe.994

Zhao, G. D., Bryan, B. A., King, D., Luo, Z. K., Wang, E. L., Bende-Michl, U., … Yu, Q. (2015). Large-scale, high-resolution agricultural systems modelling using a hybrid approach combining grid computing and parallel processing. *Environmental Modelling and Software 41*, 231–38. doi.org/10.1016/j.envsoft.2012.08.007

17

Land Use Planning as a Collective Learning Spiral: The Case of Regenerative Landscape Policy and Practice

Richard Thackway, Valerie A. Brown, David Marsh and John A. Harris

Key Points

- Landscapes and land managers are intertwining social and biophysical systems that experience their own changes over time. Past land use decisions affect current conditions, which in turn strongly affect opportunities for future use and management.

- Landscapes are complex cultural and biophysical constructs; decisions that lead to improving, rather than degrading, social, economic and environmental ideals are more likely to be achieved by using a social learning cycle.

- Recognising the interconnected changes of landscapes and natural resource management (NRM) in a form that services the needs of both planners and land managers provides an alternative approach to developing land use policies and land use plans and evaluating their outcomes.

- A social learning cycle that starts with ideals, describes the context, designs new ideas and puts them into action is familiar to both planning and management. An example of regenerative landscape planning, management and evaluation is presented below.

- Key indicators of landscape and management functions are proposed for use in this continual change process; indicators offer a basis for monitoring, evaluating and learning whether our cultural and biophysical resources are regenerative.

Introduction: The Context of Land Use Planning in Australia

Australia's Indigenous peoples did not own land or resources individually; they shared them and managed their lifestyles so that they could survive, even in the most difficult of circumstances (Arabena, 2015). For most of the time, despite Australia's uncertain climate, they lived with abundance (Gammage, 2011). The land itself, with its ancient, eroded soil base and extremes of weather conditions, was able to maintain productivity and its rich cultural living systems for over 40,000 years. European ideas of private land ownership and maximum resource extraction arrived with the First Fleet. Observations of the coastal landscape around Sydney in 1788 describe open forest and well-grassed parkland; fruit plantations and water retention systems occupied the fertile landscapes (Fitzhardinge, 1979).

Since European occupation, achievable land use decisions have been mainly judged against economic goals, with objectives based primarily on maximising yield (see Chapter 1). In current land use policy and practice, the status of the environmental resource base is either ignored or discounted. This presents an ethical question about responsibility for the landscape. A collective social learning process is required that links the spending of money and application of human capital with movement towards, or away from, desired social and environmental ideals. In a shifting climate and fast-changing human society, the outcomes of decisions in land use planning and farming practice need to be continually monitored and evaluated.

Land use planning then becomes a continual learning process for farmers and policymakers, and landscape and land managers (Figure 17.1). A farming family can set personal ideals for the family, business and farm resource base. A government may establish more broadly based, but equally meaningful, ideals for a region or the nation. Without such ideals, there is nothing against which to test decisions; ideals come first. Decision-makers for landscapes at all physical scales, and from all sector interests, need both landscape and management literacy if their decisions are to be based on a sound understanding of what is happening in specific contexts. The following 'framing questions' might be used to inform collective decision-making towards the regenerative management of Australian landscapes:

1. Developing ideals (i.e. what should be)—how do we Australians want to live on our diverse landscape, recognising that the regenerating capacity of our unique landscapes provide the basis for our physical, economic and social health?

2. Describing context (i.e. what is)—what landscape functions (energy flows, living systems, water cycles, minerals and soils and management effects), coupled with management functions (biophysical measures, socio-economic frameworks, ethical principles, aesthetic patterns and sympathetic relationships), best support the ideals for this regenerative landscape?

3. Designing collective ideas (i.e. what could be)—which old and new aspects of planning and practice would best bring about the social and landscape learning that supports this regenerating landscape?

4. Doing the collective design in action (i.e. what can be)—what lessons can be drawn from landscape and management monitoring and evaluation for the next cycle of regenerative management?

Collective Learning: The Practice

Once a group or individuals have agreed on their interest in a particular regenerative landscape, these four framing questions may be implemented as a series of steps (see Figure 17.1).

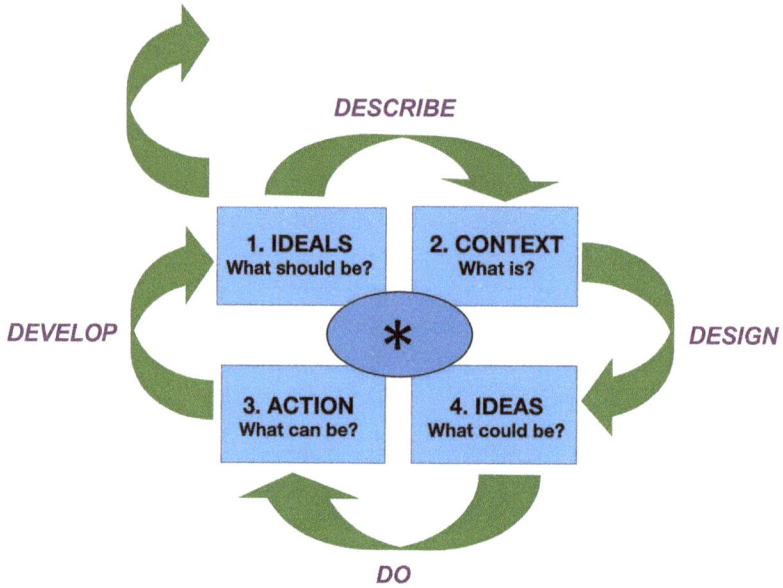

✳Regenerative Landscape Management

Figure 17.1: Collective learning cycle for regenerative landscape management.
Source: Modified from Kolb (1984).

- Step 1. Respecting difference: what should be? (developing and sharing participants' ideals)
- Step 2. Ground truthing: what is? (describing interconnected landscape and management functions)
- Step 3. Brainstorming: what could be? (designing the next program drawing on creative ideas)
- Step 4. Acting: what can be? (doing the design in partnership with established practices)

Repeating the learning cycle (Figure 17.1) at regular intervals provides a continual monitoring, spiral and evaluation system (Figure 17.2). Feedback on each step serves the needs of policy development, planning, investment, mentors, community and farmers. The diagram represents the steps of the classic, adult learning cycle that is already familiar to all these interests—that is, developing ideals of what should be occurring, documenting the context of what is occurring, designing policies and programs that could be implemented to meet the ideals and implementing the ideas that can be put into practice (Brown & Lambert, 2013).

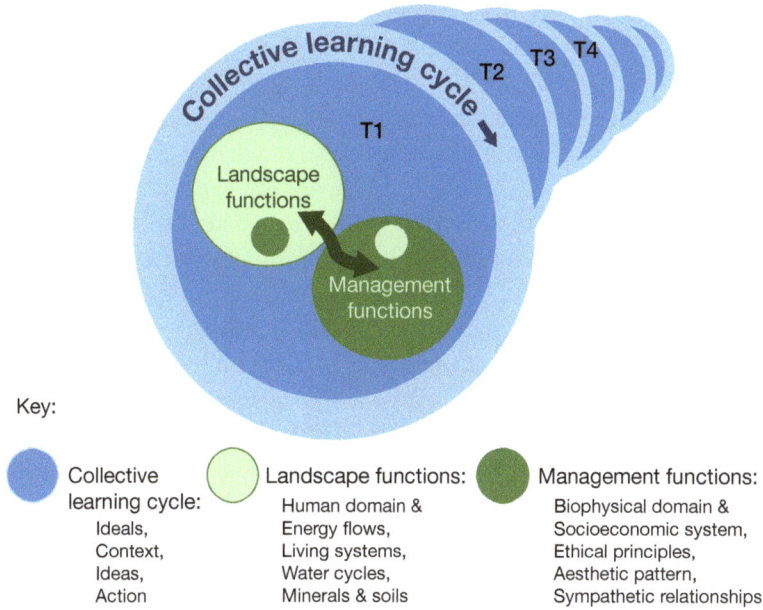

Figure 17.2: Monitoring regenerative landscape management.
Source: Authors' work.

Land Use Planning as a Collective Learning Spiral

Under current conditions for data collection, there may already exist relevant landscape and management data sets that have been collected over time. However, these collections are likely to be stored in separate compartments, frozen in time, recorded in different languages and based on incompatible units of measurement. Regenerative land use planning calls for a collaboration among the farming community, researchers, planners and innovators. This learning partnership extends to improving social health (see Chapter 11), more rigorous research (Thackway & Lesslie, 2008) and more reliable land use planning (see Chapter 4). If land use planning is to satisfy all these needs, there must be a firm basis for collecting the full range of existing evidence for both biophysical and social landscapes.

The collective learning spiral—developing ideals, describing the context, designing with new ideas and putting these into practice—has long been shown to apply to individual adult learning (Kolb, 1984). Over the past 30 years, the continual learning cycle (Figures 17.1 and 17.2) has been developed as a social learning cycle, evaluating and responding to transformational physical and social change (Brown, 1996; Brown & Lambert, 2013). When the aim is to enable change rather than maintain the status quo or revert to the past, the application of the learning cycle is very different. This is the case in supporting regenerative landscapes. Existing NRM programs traditionally begin by recalling 'the facts' and describing *what is presently occurring* as the basis for fresh decisions (Step 2, Figure 17.1): this is a mistake. Programs for change traditionally start with brainstorming new ideas—*what could ideally be occurring* (i.e. what should be done), followed by the continual testing of innovative practices (Step 1, Figure 17.1). In contrast, conventional planning regimes and on-ground practice may focus on *what can be* done *in practical terms*, which amounts to a continuation of current practices. In each of these cases, unfortunately, the response to a social, environmental or economic challenge is to continue the existing way of doing things.

Ideally, a change management process should develop fresh ideals for a different future. Where the aim is maximising movement towards an ideal, instead of defending existing practice, adoption of continual learning cycles is required in the sequence of learning, as shown in Figure 17.2 (see Brown & Lambert, 2013). Each cycle of the learning spiral starts by refreshing ideals in light of new experiences. The lessons learned through putting ideas into practice may change them in small or large ways, which results in the cycle beginning anew.

Land use planning programs that have applied this cycle include the Australian Decade of Landcare, the World Health Organization's Healthy Cities, the University of Queensland's International Water Centre, Future Earth's Transformations to Sustainability and The Australian National University's Local Sustainability Project. At the farming scale, emergent programs utilising this cycle include Holistic Management, Resource Consulting Service, Natural Sequence Farming, Prograze, Landscan and the Alliance for Regenerative Landscapes and Social Health.

Each cycle of the learning spiral for regenerative landscapes is equally relevant to leaders in the field, innovative farmers, experts, influential organisations and creative thinkers (Brown, Harris & Russell, 2010).

Since each of these interests has their own mode of learning and sphere of power and control, each will approach land use planning from a different perspective. Farmers and farm managers are often sole operators. Organisational interests will work within the web of their own networks. Experts operate within the frameworks and languages of their own communities of practice. Creative thinkers can come from 'left field'. Regardless of these differences, the collective learning cycle (Figure 17.1) acts as a framework for generating the ideals, context, ideas and actions of a continuous exchange process that works for all.

Step 1. Collective Learning Cycle: Developing Ideals for Regenerative Landscapes

Developing ideals is the first step in the land use planning process. The ideal of land use planning as a partnership with continually regenerating landscapes is gathering momentum throughout Australia and, indeed, the world (Mutizwa, 2015). The four authors of this paper—a regenerative farmer, an ecologist, a land use planner and a change management practitioner—described their contributions to each step of the learning cycle below.

A Farmer's Ideals

- Socio-economic: 'Living on our land in a secure family that values friendship, cooperation, humour and learning; producing a profit from livestock and plants, and [having] enough time for recreation.'

- Environmental: 'A diverse landscape with pasture year round, an efficient water cycle, soil life cycling minerals, and an increasing sunlight harvest, and management by hard work and collaboration with the local community.'

An Ecologist's Ideals

- Socio-economic: 'Personal opportunities for experiential learning, mental reflection and sympathetic curiosity, inspired by planetary interrelationships which maintain self-organising systems sustainably.'

- Environmental: 'In all species, mutually supportive communities, respect for individual differences, an ethic of justice and fairness, aesthetic patterns intrinsic to regenerative landscapes and feelings of empathy for all forms of life.'

A Planner's Ideals

- Socio-economic: 'Land use planning built on sound data for management decisions, requiring continual monitoring of landscape and management functions towards or away from ideals.'

- Environmental: 'Land use that allows for capital increase of landscape functions: energy flows, biota, water cycles, minerals and soils and management.'

A Change Manager's Ideals

- Socio-economic: 'Collective learning between all the interests in a region working collaboratively to change existing land use planning from mechanical to regenerative agriculture and from capital growth to expanding common pool resources.'

- Environmental: 'Expanding literacy in the self-organising capacity of all environmental functions and their interrelationships with management functions.'

Step 2. Collective Learning Cycle: Describing the Context of the Regenerating Landscape

After hearing each other's ideals in Step 1, the learning moves to ground-testing those ideals in the context in which they will become operational. This seems a tall order, especially when considered alongside current practices, for which participants collect their own data, for their own purposes, in their own language. Close examination reveals that there is a single set of functions involved, and that most (if not all) of the functions are already the concern of data collections and research programs. This step in the spiral is strength-based, rather than problem-based. The question 'what is?' can be answered by sharing observations under the three main headings in Figure 17.2.

Landscape functions are shaped by five interrelated biophysical processes (Massy, 2017), namely:

- energy flows (i.e. movement of energy from the sun through a landscape)

- living systems (i.e. increasing complexity of ecological systems)

- water cycles (i.e. movement of water from the atmosphere to the landscape and back)

- minerals and soils (i.e. movement of minerals and nutrients through living systems)
- management domain (i.e. impact of the current management system on the landscape).

Management functions are shaped by five interrelated ways of understanding the issues (Brown & Harris, 2014):

- biophysical domain created by the landscape functions of energy, life, water and soils
- socio-economic frameworks that shape livelihoods (e.g. democracy, capitalism and productivity)
- ethical principles guiding landscape decisions (e.g. partnership, stewardship and land ownership)
- aesthetic patterns admired or rejected (e.g. pasture, crops, weeds, waterways and native forests)
- sympathetic relationships that shape personal actions (e.g. family, friendships and fauna).

Example of Collective Learning Spiral Step 2

An example of the dual management and biophysical monitoring processes are the observations recorded for the grazing system known as 'Chandler Paddock' at Severn Park in the Monaro Region, New South Wales. The manager of Severn Park described the management functions he would draw on in planning and managing a regenerative landscape thus (Massy, 2017):

- Biophysical observations: 'From 65 regenerative agriculturalists and over 2000 generations of Indigenous Australians, the conclusion is that it is only via close ecological endeavour coupled with transformational social learning that we can transform ourselves and our behaviours so as to regenerate and sustain our living and non-living support systems for the future.'
- Socio-economic framework: shifting from the ruling mechanical, neo-liberal landscape management framework to the emerging regenerative neo-organic.
- Ethical principles: reinstating moral imperatives of collaboration with the landscape itself and among a community of regenerative farmers.

- Aesthetic patterns: the mosaic pattern of patchwork burning, diversity of plant species in a regenerating paddock.
- Sympathetic relationships: intimacy between people and their landscapes.

The basis for this description of management functions for Chandler Paddock is the set of ideals outlined under Step 1. The human interests that determine the management regime have shared their biophysical, socio-economic, ethical, aesthetic and sympathetic ways of understanding what they seek in their management in Step 1 (see Figure 17.3). Each of these is reviewed for their status in relation to the particular place.

The basis for the description of the landscape functions in Step 2 is the vegetation, assets, states and transitions (VAST) framework ecological analysis. VAST has been developed to diagnose, account for, monitor and report on the interactions between the biophysical effects of land management on transforming landscapes over time (Thackway & Lesslie, 2008). VAST classifies and maps vegetative land cover into 'condition classes' that reflect the energy, living systems, water status, mineral functions and management functions of a particular landscape. It follows that the categories of indicators (structure, composition and function) can be the same for all landscapes, while the actual indicators are tied to place (Thackway & Specht, 2015).

For example, Chandler Paddock is an area of 182 hectares comprising rolling and hilly terrain, located at −36°26'23.4600"S, 148°55'42.0000"E. After 125 years of maximum human, technical and chemical resource application, all indicators of landscape functions had fallen considerably from a reference in 1750. With the shift to regenerative farming in 2000, all indicators began a return towards the reference state (80–100 per cent).

Six phases were involved in the transformation of Chandler Paddock (Figure 17.3). Phase 1 describes the reference state. Phases 2–5 involved intensive grazing management with commensurate declines in the structure, composition and function of the site. The inclusive process, involving Steps 1–4 above, was used to initiate regenerative landscape management practices, resulting in Phase 5. A change management process, involving collective social learning, was used to initiate a new phase of regenerative landscape management (Phase 6). The observed increases in regenerative capacity, vegetation status and energy, soil water and carbon storage continue to accrue.

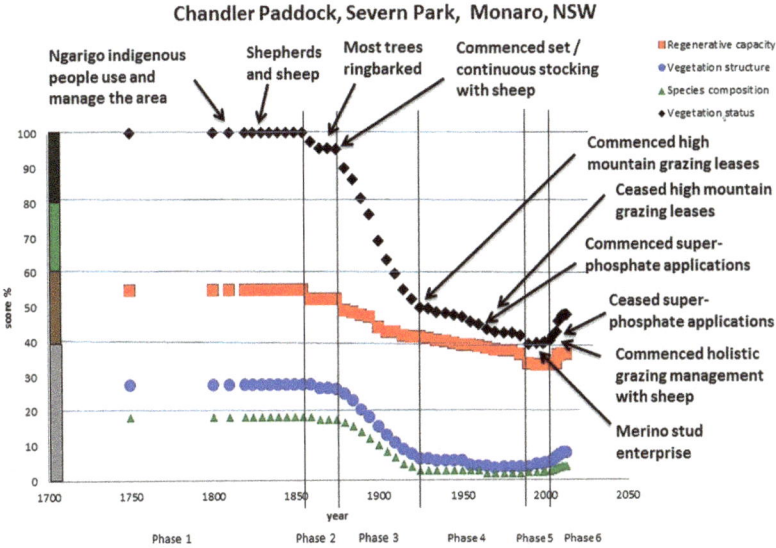

Figure 17.3. Six phases of the results of land use planning.
Source: Modified from Yapp & Thackway (2015).

Step 3. Collective Learning Cycle: Ideas for Regenerative Landscape Management

After determining the ideals for their regenerative landscapes, and identifying the current supports and inhibitors of those ideals, the management cluster is in a good position to generate creative ideas that work over the long term.

A farmer's report:

> We are turning the previous intensive farming paradigm on its head. We have seen diversity increase naturally, a key step towards achieving a sustainable social, economic and environmental resource base ideal. Previous energy intensive inputs were deemed detrimental and unnecessary. We now try to manage with contemporary sunlight energy. The effect of managing holistically on our family and the day-to-day management has been appreciating the wonder of the natural world, and knowing we can approach any type of climate variability with confidence.

A planner's report:

> Ideas that have become part of regenerative landscape management include the use of the continual learning/planning cycle [see Figure 17.2] to initiate regenerative landscape outcomes [see Figure 17.3]; an acceptance of the richness of diversity in the landscape and among landscape decision-makers; and recognition that constant change is part of the dynamic landscape and management interrelationship, and requires constant monitoring, reporting and evaluation.

Step 4. Collective Learning Cycle: Putting the Ideas into Regenerative Physical and Social Practice

Once the ideas are on the table (Step 3, Figure 17.1 and Phase 5, Figure 17.3), the decision-makers can determine which ideas are the most promising and the most practical. The current dominant landscape monitoring approach separates the biophysical and management functions, resulting in poor integration of information and suboptimal social, economic and environmental outcomes. In contrast, when decision-makers (i.e. land use policy and planning, as well as land managers) recognise that they are operating within the same regenerating landscape and that they share responsibility for working with the same set of management and landscape functions, they can represent these functions as interconnected indicators (Figure 17.4). When continually monitored for informed decision-making (Figure 17.2), these indicators improve the potential for continual social and biophysical learning.

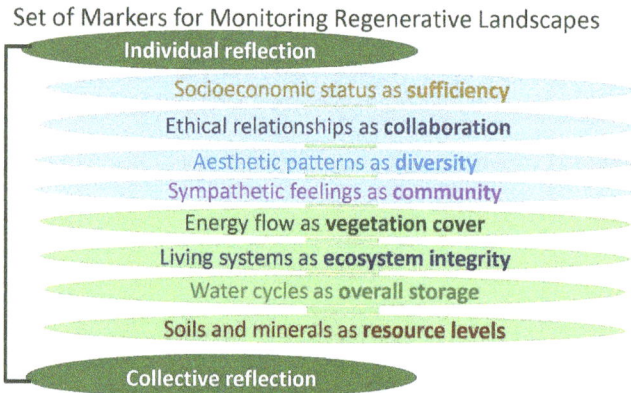

Set of Markers for Monitoring Regenerative Landscapes

Individual reflection

Socioeconomic status as **sufficiency**

Ethical relationships as **collaboration**

Aesthetic patterns as **diversity**

Sympathetic feelings as **community**

Energy flow as **vegetation cover**

Living systems as **ecosystem integrity**

Water cycles as **overall storage**

Soils and minerals as **resource levels**

Collective reflection

Figure 17.4: Eight indicators of landscape and management functions that are applied in Figures 17.1 and 17.2.

Source: Authors' work.

The indicators in Figure 17.4 bring together landscape and management functions for the full process of sustainable regenerative management to be monitored over time in a collaborative, appreciative process that enhances summative evaluation and learning.

The following two testimonies, which document the process of putting the ideas into regenerative physical and social practice, are informative.

A farmer's report:

> We have [learned] to match our stocking rate to the carrying capacity of the landscape as it changes dynamically. This has given us confidence in knowing when to reduce stock numbers instead of hand feeding stock. Our costs have plummeted and our business is carrying a lot less debt than before. By matching stocking rate to carrying capacity we maintained ground cover during nine years of drought 2002–2010, and did not spend any money feeding stock, a huge saving. Insects, reptiles and local tree species appear in abundance. The living community has become more diverse, its natural tendency.

A planner's report:

> In December 2015 there were 45 districts in New South Wales with a total of 470 regenerative landscape farmers; 109 in Western Australia, 76 in Victoria, and 7 in Tasmania involved with holistic planning (Australian Holistic Management Certified Educators). Estimations from other groups is over 700 (Marsh, David, pers. comm.). This equals the 1500 Australian farming households involved in Landcare at its peak (Brown, 1996).

The set of indicators (Figure 17.4) brings together the many ways of learning (Brown & Harris, 2014) for each of the four learning steps (Figures 17.1 and 17.2). In practical terms, most participants will only evaluate Step 4 (Figure 17.1), whereas we propose that participants use all eight indicators to continuously improve learning in all four steps.

Conclusions and Recommendations

This chapter outlines a social learning framework for the development of monitoring and evaluation processes to transform current land management practices from a productivity-driven approach to one based on the regeneration of land and social capital. Landscapes and land managers are intertwining social and biophysical systems that experience their own changes over time. Past land use decisions affect current conditions, which, in turn, affect opportunities for future use and management. Landscapes are transformed by land use policies and planning that affect how the land is managed.

In complex cultural and biophysical landscapes, decisions that lead to improving, rather than degrading social, economic and environmental ideals are more likely to be achieved by using the learning cycle, in which every step is multidimensional. The magnitude of change in land management over time, and the impacts on ecosystem transformation, will determine the extent to which a landscape is regenerative. The degree to which a landscape has been changed, and an estimate of the potential for that landscape to regenerate naturally, or to be restored, involves an initial assessment and interpretation of these criteria and their associated indicators (Figure 17.3).

The degree of modification from the natural reference state will determine how likely a land manager is to be successful in improving all, or some, of the landscape's structure, composition and function. The regenerative capacity of a landscape and its component ecosystems will be determined by the degree to which an ecosystem's structure, composition and function has previously been modified, removed or replaced, as well as the type of management that is practical and feasible within cultural and biophysical constraints.

Regular monitoring and evaluation using a set of indicators of landscape and management functions, implemented as part of the continual learning cycle, are likely to improve social, economic and environmental outcomes. The development of national land use policy and planning decisions should consider the tools needed to assist decision-makers to assess regenerative landscapes and social health. That knowledge can then be used to promote and encourage the wider adoption of land management practices that, in turn, can promote improved social health.

References

Arabena, K. A. (2015). *Becoming Indigenous to the universe*. Melbourne, VIC: Academic Press.

Brown V. A. (Ed.). (1996). *Landcare languages: Talking to each other about living with the land.* Canberra, ACT: Department of Primary Industries.

Brown V. A. & Harris, J. A. (2014). *The human capacity for transformational change: Harnessing the collective mind.* London, UK: Routledge.

Brown, V. A., Harris, J. A. & Russell, J. Y. (2010). (Eds.). *Tackling wicked problems: Through the transdisciplinary imagination.* London, UK: Earthscan.

Brown, V. A. & Lambert, J. A. (2013). *Collective learning for transformational change: A guide to collaborative action.* London, UK: Routledge.

Fitzhardinge, L. F. (1979). *A reprint of Sydney's first four years: A narrative of the expedition to Botany Bay and a complete account of the settlement of Port Jackson by Captain Watkin Tench, British Marine Officer.* Sydney, NSW: Library of Australian History.

Gammage, B. (2011). *The biggest estate on earth: On how Aborigines made Australia.* Sydney, NSW: Allen & Unwin.

Kolb, D. A. (1984). *Experiential learning: Experience as the source of learning and development.* Englewood Cliffs, NJ: Prentice Hall.

Massy, C. (2017). *Call of the Reed Warbler: A new agriculture – a new earth.* St Lucia: University of Queensland Press.

Mutizwa, M. (2015). *Developmental work research: A tool for enabling agricultural innovation.* Waginingen: Waginingen Academic Publishers (online). doi.org/10.3920/978-90-8686-819-3

Thackway, R. & Lesslie, R. (2008). Describing and mapping human-induced vegetation change in the Australian landscape. *Environmental Management 42,* 572–90. doi.org/10.1007/s00267-008-9131-5.

Thackway, R. & Specht, A. (2015). Synthesising the effects of land use on natural and managed landscapes. *Science of the Total Environment 526,* 136–52. doi.org/10.1016/j.scitotenv.2015.04.070.

Yapp, G. A. & Thackway, R. (2015). Responding to change—criteria and indicators for managing the transformation of vegetated landscapes to maintain or restore ecosystem diversity. In J. A. Blanco (Ed.), *Biodiversity in ecosystems—linking structure and function.* London UK: InTech. Retrieved from www.intechopen.com/books/biodiversity-in-ecosystems-linking-structure-and-function/responding-to-change-criteria-and-indicators-for-managing-the-transformation-of-vegetated-landscapes

Appendix: Index of publications by Dr Rob Lesslie

This appendix serves as an index to the three volumes that comprise a complete collection of the published scientific papers and book chapters of Dr Rob Lesslie (1957–2014).

- Volume 1: Wilderness publications, 1981–2016
- Volume 2: Land use publications, 2000–2013
- Volume 3: Multi-criteria analysis publications, 2005–2014

These hardcover, cloth-bound volumes have been lodged as a teaching resource with the Fenner School of Environment and Society, The Australian National University.

Volume 1, Wilderness Publications, 1981–2016

Lesslie, R. (2016). The wilderness continuum concept and its application in Australia: lessons for modern conservation. In S. J. Carver & S. Fritiz (Eds.), *Mapping Wilderness: Spatial Methods and Applications in Mapping and Modelling Wild Landscapes* (pp. 3–33). Netherlands: Springer.

Lesslie, R. (2012). Mapping our priorities—innovation in spatial decision support. In P. Figgis, J. Fitzsimons, & J. Irving (Eds.), *Innovation for 21st Century Conservation* (pp. 46–153). Sydney, NSW: Australian Committee for IUCN.

Lesslie, R., Thackway, R., & Smith, J. (2010). *A national-level vegetation assets, states and transitions (VAST) dataset for Australia* (version 2.0), Bureau of Rural Sciences, Canberra.

Thackway, R. & Lesslie, R. (2008). Describing and mapping human-induced vegetation change in the Australian landscape. *Environmental Management 42*, 572–90.

Mackey, B. G., Soulé, M. E., Nix, H. A., Recher, H. F., Lesslie R. G., Williams, J. E., Woinarski, J. C. Z. R., Hobbs, J. & Possingham, H. P. (2007). Towards a scientific framework for the WildCountry project. In J. Wu & R. J. Hobbs (Eds.), *Key Topics and Perspectives in Landscape Ecology* (pp. 2–208). Cambridge University Press.

Thackway, R. & Lesslie, R. (2006). Reporting vegetation condition using the vegetation assets, states and transitions (VAST) framework. *Ecological Management and Restoration 7*, S53–S62.

Thackway, R. & Lesslie, R. (2005). *Vegetation Assets, States and Transitions: Accounting for Vegetation Condition in the Australian Landscape* (Technical Report). Canberra, ACT: Bureau of Rural Sciences.

Lindenmayer, D. B., Cunningham, R. B., Donnelly, C. F. & Lesslie, R. G. (2002). On the use of landscape surrogates as ecological indicators in fragmented forests. *Forest Ecology and Managemen, 159*(3), 203–16.

Lesslie, R. G. (2001). Landscape classification and strategic assessment for conservation: an analysis of native cover loss in far south-east Australia. *Biodiversity and Conservation 10*(3), 427–42.

Kapos, V., Lysenko, I. & Lesslie, R. (2000). *Assessing Forest Integrity and Naturalness in Relation to Biodiversity.* Cambridge, UK: UNEP-WCMC.

Lesslie, R. G. (2000). Terrestrial wilderness (map). In B. Groombridge & M. D. Jenkins (Eds.), *Global Biodiversity: Earth's Living Resources in the 21st Century* (p. 1). Cambridge, UK: World Conservation Monitoring Centre.

Environmental Research and Assessment Pty Ltd. (1999). *Wilderness in Western, Eastern and Southern South Australia,* (Lesslie, R. author). Three volumes. A report to the Australian and World Heritage Group, Environment Australia, Canberra, ACT.

Lesslie, R. G. (1999). *The Size of Wilderness.* Sydney, NSW: New South Wales National Parks and Wildlife Service.

Lesslie, R. G. (1998). *Interaction Between Human Activity and Forest Biota at Landscape Scales.* Report to the Forest Taskforce, Environment Australia, Canberra.

Mackey, B. G., Lesslie, R. G., Lindenmayer, D. B. & Nix, H. A. (1998). Wilderness and its place in nature conservation in Australia. *Pacific Conservation Biology 4*(3), 182–5.

Mackey, B. G., Lesslie, R. G. Lindenmayer, D. B., Nix, H. A. & Incoll, R. D. (1998). *The Role of Wilderness in Nature Conservation.* Report to the Australian and World Heritage Group, Environment Australia.

Tickle, P., Hafner, S., Lesslie, R., Lindenmayer, D., McAlpine, C. Mackey, B., Norman, P. & Phinn, S. (1998). *Montreal Indicator 1.1e Fragmentation of Forest Types: Identification of Research Priorities.* Report to the Forest and Wood Products Research and Development Corporation, Canberra, ACT.

Lesslie, R. G. (1997). *A Spatial Analysis of Human Interference in Terrestrial Environments at Landscape Scale.* Unpublished PhD thesis, Department of Geography, The Australian National University, Canberra, ACT.

Lesslie, R. G. (1995). *Assessing Environmental Integrity and the Effects of Human Activity in the Landscape: A Review.* Report to the Australian Heritage Commission, Canberra, ACT.

Lesslie, R. (1995). *National Wilderness Inventory: Human Interference in the Natural Landscape.* Report to the Australian Heritage Commission, Canberra, ACT.

Lesslie, R. G. (1995). Wilderness assessment and modelling in Australia using GIS-based modelling techniques. In R. B. Singh (Ed.), *Global Environmental Change—Perspectives of Remote Sensing and Geographic Information Systems* (pp. 7–118). New Delhi: Oxford and IBH Publishing.

Lesslie, R. G. & Fitzgerald, N. (1995). *Australian Offshore Islands Wilderness Quality Assessment.* Report to Anutech Pty Ltd.

Lesslie, R. G. & Maslen, M. (1995). *National Wilderness Inventory: Handbook of Procedures, Content and Usage* (2nd ed.). Canberra, ACT: Australian Government Publishing Service.

Lesslie, R. (1994). The Australian National Wilderness Inventory: Wildland survey and assessment in Australia. In *Wilderness: The Spirit Lives* (pp. 4–97). Proceedings of the 6th US National Wilderness Conference, Santa Fe, USA, November 1994.

Lesslie, R. (1994). The National Wilderness Inventory: Wild-land identification, assessment, and monitoring in Australia. In J. Hendee & V. Martin (Eds.), *International Wilderness Allocation, Management and Research* (pp. 1–37). Proceedings of a symposium during the 5th World Wilderness Congress, Tromso, Norway, September 1993. Paris: World Heritage Centre.

Lesslie, R., Maslen, M. & Taylor, D. (1994). The National Wilderness Inventory. In W. Barton (Ed.), *Wilderness—The Future* (pp. 9–41). Sydney, NSW: Envirobook, Sydney.

Lesslie, R. G. (1993). *Towards a National System of Forest Reserves: Wilderness Issues.* Report to the Department of the Arts, Sport, the Environment and Territories, Canberra, ACT.

Lesslie, R. G. & Taylor, D. (1994). *National Wilderness Inventory: The Base-line Survey (vols 1 and 2).* Report to the Australian Heritage Commission.

Lesslie, R. G., Taylor, D. & Maslen, M. (1993). *National Wilderness Inventory: Handbook of Principles, Procedures and Usage.* Canberra, ACT: Australian Government Publishing Service.

Lesslie, R. G., Abrahams, H. & Maslen, M. (1992). *National Wilderness Inventory Stage III: Wilderness quality on Cape York Peninsula.* Canberra, ACT: Australian Heritage Commission.

Lesslie, R. G. (1991). Wilderness survey and evaluation in Australia. *Australian Geographer 22*, 35–43.

Lesslie, R. G. (1991). *An Assessment of Wilderness Values for Western Tasmania National Estate Extension Areas.* Report to the Australian Heritage Commission, Canberra, ACT.

Fuller, A., Leadbeater, P., Leaver, B., Lesslie, R., & Ware, R. (1990). *Wilderness Protection Act.* White Paper. South Australian Department of Environment and Planning, Adelaide, SA.

Lesslie, R.G. (1990). Procedures for the evaluation and management of wilderness. In J. C. Noble, P. J. Joss & G. K. Jones (Eds.), *The Mallee Lands: A Conservation Perspective.* Melbourne, VIC: CSIRO Publishing.

Lesslie, R. G., Maslen, M. Canty, D. & Goodwins, D. (1990). *Wilderness Quality on Kangaroo Island.* Report to the Department of Environment and Planning, South Australia and the Australian Heritage Commission.

Mackey, B. G., Lesslie, R. G. & Dovey, L. (1990). *Biophysical Naturalness Assessment of the South-East Forests of New South Wales.* Report to the National Parks and Wildlife Service, New South Wales.

Lesslie, R. G. (1989). *Wilderness Quality and the Distribution of Old Growth Eucalypt Forest in Tasmania.* Report to the World Heritage Unit, Department of Arts, Sport, the Environment, Tourism and Territories, Canberra, ACT.

Lesslie, R. G., Mackey, B. G. & Preece, K. M. (1988). A computer-based method for wilderness evaluation. *Environmental Conservation 15*(3), 225–32.

Lesslie, R. G., Mackey, B. G. & Schulmeister, J. (1988). *Wilderness Quality in Tasmania.* Canberra, ACT: Australian Heritage Commission.

Lesslie R. G., Mackey B. G. & Schulmeister, J. (1988). *National Wilderness Inventory Wilderness quality in Victoria.* A report to the Australian Heritage Commission. Commonwealth of Australia, Canberra, ACT.

Lesslie, R. G. & Stein, J. A. (1988). Analysis of the Report of the Commission of Inquiry into the Lemonthyme and Southern Forests with Respect to Wilderness. Report to the Australian Heritage Commission, Canberra, ACT.

Lesslie, R. G., Mackey, B. G. & Preece, K. M. (1987). *A Computer-Based Methodology for the Survey of Wilderness in Australia.* Report to the Australian Heritage Commission, Canberra, ACT.

Preece, K. M. & Lesslie, R. G. (1987). *A Survey of Wilderness Quality in Victoria.* Ministry for Planning and Environment Victoria and Australian Heritage Commission, Canberra, ACT.

Lesslie, R. G. & Taylor, S. G. (1985). The wilderness continuum concept and its implications for Australian wilderness preservation policy. *Biological Conservation 32*, 309–33.

Lesslie, R. G. & Taylor, S. G. (1983). *Wilderness in South Australia:* Occasional Paper No. 1, Centre for Environmental Studies, University of Adelaide, Adelaide, SA.

Lesslie, R. & Taylor, S. G. (1983). South Australia: A different kind of wilderness. *Habitat Australia 11*(6), 28–30.

Lesslie, R. G. (1982). *The Conservation of Wilderness in the Ninety Mile Desert and Coffin Bay Peninsula, South Australia.* Report to the South Australian Reserves Advisory Committee, Adelaide, SA.

Lesslie, R. G. (1981). *Wilderness: An Inventory Methodology and Preliminary Survey of South Australia.* Unpublished Master of Environmental Studies thesis, Centre for Resource and Environmental Studies, University of Adelaide, Adelaide, SA.

Volume 2, Land Use Publications, 2000–2013

Clancy, T. F. & Lesslie, R. G. (2013). *A Scoping Assessment for a National Research Centre Addressing Land Use and Food Security Issues.* Australian Bureau of Agricultural and Resource Economics and Sciences. Report prepared for International Agricultural Cooperation Program, Trade and Market Access Division, Department of Agriculture, Canberra, ACT.

Lesslie, R. & Mewett, J. (2013). *Land Use and Management: The Australian Context.* Research Report 13.1, Australian Bureau of Agricultural and Resource Economics and Sciences, Canberra, ACT.

Mewett, J., Paplinska, J., Kelley, G., Lesslie, R., Pritchard, P. & Atyeo, C. (2013). *Towards National Reporting on Agricultural Land Use Change in Australia* (Technical Report 13.06). Australian Bureau of Agricultural and Resource Economics and Sciences, Canberra, ACT.

Lesslie, R. (2012). Mapping our priorities—innovation in spatial decision support. In P. Figgis, J. Fitzsimons, J. Irving (Eds.), *Innovation for 21st Century Conservation* (pp. 46–153). Sydney, NSW: Australian Committee for IUCN.

Australian Bureau of Agricultural and Resource Economics and Sciences. (2011). *Multi-criteria Analysis Shell for Spatial Decision Support (MCAS-S) User Guide.* Canberra, ACT: Australian Bureau of Agricultural and Resource Economics and Sciences.

Australian Bureau of Agricultural and Resource Economics and Sciences. (2011). *Guidelines for Land Use Mapping in Australia: Principles, Procedures and Definitions* (4th ed.). Canberra, ACT: Australian Bureau of Agricultural and Resource Economics and Sciences.

Lesslie, R., Mewett, J. & Walcott, J. (2011). Landscapes in transition: Tracking land use change, *Science and Economic Insights 2.2.* Canberra, ACT: Australian Bureau of Agricultural and Resource Economics and Sciences.

Australian Collaborative Land Use and Management Program. (2010). *Land Use and Land Management Information for Australia: Workplan of the Australian Collaborative Land Use and Management Program.* Canberra, ACT: Australian Bureau of Agricultural and Resource Economics and Sciences.

Lesslie, R., Thackway, R., & Smith, J. (2010). *A national-level vegetation assets, states and transitions (VAST) dataset for Australia* (version 2.0). Canberra, ACT: Bureau of Rural Sciences.

Stenekes, N., Kancans, R., Randall, L., Lesslie, R., Stayner, R., Reeve, I. & Coleman, M. (2010). *Indicators of Community Vulnerability and Adaptive Capacity Across the Murray Darling Basin—A Focus on Irrigation in Agriculture.* Report to the Murray Darling Basin Authority. Canberra, ACT: Bureau of Rural Sciences; Armidale, NSW: Institute for Rural Futures, University of New England.

Cotsell, P., Gale, K., Hajkowicz, S., Lesslie, R., Marshall, N. & Randall, L. (2009). Use of a multiple criteria analysis (MCA) process to inform Reef Rescue regional allocations. In S. Hogan & S. Long (Eds.), *Proceedings of the 2009 Marine and Tropical Sciences Research Facility Annual Conference,* 28–30 April 2009. Reef and Rainforest Research Centre Limited, Cairns.

Lesslie, R., Smith, J. & Rickards, J. (2009). Agricultural land use mapping and decision-support in Australia. In Y. Chen, Y. He, & Q. Yu (Eds.), *Agricultural Land Use and its Effect in APEC Member Economies* (pp. 1–19). APEC Agricultural Technical Cooperation Working Group, Beijing.

Barson, M., Lesslie, R., Smith, J. & Stewart, J. (2008). Developing land cover and land use data sets for the Australian continent—a collaborative approach. In J. C. Campbell, K. B. Jones, J. H. Smith & M. T. Koeppe (Eds.), *North America Land Cover Summit* (pp. 5–74). Washington, DC: Association of American Geographers.

Byron, I. & Lesslie, R. (2008). Spatial methodologies for integrating social and biophysical data at regional or catchment scale. In R. j. Aspinall & M. J. Hill (Eds.), *Land Use Change: Science, Policy and Management* (pp. 3–62). Boca Raton, FL: Taylor Francis Group.

Lesslie, R. G., Barson, M. M. & Randall, L. A. (2008). Land use mapping. In N. J. McKenzie, M. J. Grundy R. Webster & A. J. Ringrose-Voase (Eds.), *Guidelines for Surveying Soil and Land Resources* (pp. 43–155). Collingwood, VIC: CSIRO Publishing.

Lesslie, R. & Cresswell, H. (2008). Mapping priorities: Planning re-vegetation in southern NSW using a new decision-support tool. *Thinking Bush 7,* 30–3.

Lesslie, R. G., Hill, M. J., Hill, P., Cresswell, H. P. & Dawson, S. (2008). The application of a simple spatial multi-criteria analysis shell (MCAS-S) to natural resource management decision making. In C. Pettit, W. Cartwright, I. Bishop, K. Lowell, D. Pullar & D. Duncan (Eds.), *Landscape Analysis and Visualisation* (pp. 3–95). Lecturer Notes in Geoinformation and Cartography Series, Springer, Berlin.

Thackway, R. & Lesslie, R. (2008). Describing and mapping human-induced vegetation change in the Australian landscape. *Environmental Management 42*, 572–90.

Smith, J., Atyeo, C., Thackway, R. & Lesslie, R. (2007). Profiling Australia's vegetation—its uses and values. In *Veg Futures*, Conference in the Field, 19–23 March 2006, Albury-Wodonga. Greening Australia, Canberra, ACT.

Thackway, R., Frakes, I. & Lesslie, R. (2007). Reporting trends in Vegetation Assets, States and Transitions at the farm level—a southern tablelands case study. In *Veg Futures,* Conference in the Field, 19–23 March 2006, Albury-Wodonga. Greening Australia, Canberra, ACT.

Zhang, K., Liu, N., Fu, Z., Lesslie, R. & Caelli, T. (2007). Discovering prediction model for environmental distribution maps. *11th Pacific-Asia Conference on Knowledge Discovery and Data Mining* (PAKDD'2007). CD-ROM.

Buang, N., Liu, N., Caelli, T., Lesslie, R. & Hill, M. J. (2006). Discovering knowledge from distribution maps using Bayesian networks. *Proceedings of Fifth Australasian Data Mining Conference* (AusDM2006) (pp. 9–74). Sydney.

Hill, M. J., Lesslie, R., Donohue, R., Houlder, P., Holloway, J. & Smith, J. (2006). Multi-criteria assessment of tensions in resource use at continental scale: A proof of concept with Australian Rangelands. *Environmental Management 37*(5), 712–31.

Lesslie, R., Barson, M. & Smith, J. (2006). Land use information for integrated natural resources management—a coordinated national mapping program for Australia. *Journal of Land Use Science 1*(1), 45–62.

Lesslie, R., Barson, Smith, J., Randall, L. & Bordas, V. (2006). *Co-ordinated Land Use Mapping for Australia: Information for Landscape Solutions.* Canberra, ACT: Bureau of Rural Sciences, Australian Government Department of Agriculture, Fisheries and Forestry.

Lesslie, R., Hill, M., Woldendorp, G., Dawson, S. & Smith, J. (2006). *Towards Sustainability for Australia's Rangelands: Analysing the Options.* Canberra, ACT: Bureau of Rural Sciences.

Thackway, R. & Lesslie, R. (2006). Reporting vegetation condition using the vegetation assets, states and transitions (VAST) framework. *Ecological Management and Restoration 7*, S53–S62. doi.org/10.1111/j.1442-8903.2006.00292.x.

Donohue, R., Hill, M., Holloway, J., Houlder, P., Lesslie, R., Smith, J. & Thackway, R. (2005). *Australia's Rangelands: An Analysis of Natural Resources, Patterns of Use and Community Assets.* Report to the Natural Resource Management Business Unit, Department of Agriculture, Fisheries and Forestry, Bureau of Rural Sciences, Canberra, ACT.

Hill, M. J., Lesslie, R., Barry, A. & Barry, S. (2005). A simple, portable, spatial multi-criteria analysis shell—MCAS-S. In A. Zerger & R. M. Argent (Eds.), *MODSIM 2005 International Congress on Modelling and Simulation*, December. Modelling and Simulation Society of Australia and New Zealand, Melbourne.

Thackway, R. & Lesslie, R. (2005). *Vegetation Assets, States and Transitions: Accounting for Vegetation Condition in the Australian Landscape* (Technical Report). Canberra, ACT: Bureau of Rural Sciences.

Aslin, H., Kelson, S., Smith, J. & Lesslie, R. (2004). *Peri-Urban Landholders and Bio-Security Issues – A Scoping Study.* Canberra, ACT: Bureau of Rural Sciences.

Barson M. & Lesslie R. (2004). Land management practices—why they are important and how we know this. In *National workshop on Land Management Practices: Information Priorities, Classification and Mapping,* Bureau of Rural Sciences, Canberra, ACT.

Lesslie, R., Barson, M., Bordas, V., Randall, L. & Ritman, K. (2003). Land use mapping at catchment scale: information for catchment solutions. *Science for Decision Makers,* June. Bureau of Rural Sciences, Canberra, ACT.

Lesslie, R., Barson, M. & Randall, L. (2003). *Land Use Mapping in the Goulburn-Broken, Upper-Billabong Creek and Condamine Catchments* (Technical Report). Canberra, ACT: Murray–Darling Basin Commission.

Bordas, V. & Lesslie, R. (2002). *Land Use Mapping of the Snowy Catchment in NSW.* Report to the Snowy River Shire and the Natural Heritage Trust. Canberra, ACT: Bureau of Rural Sciences.

Bureau of Rural Sciences (2002). *Land Use Mapping at Catchment Scales: Principles, Procedures and Definitions* (2nd ed.). Canberra, ACT: Bureau of Rural Sciences.

Lesslie, R., Barson, M. & Randall, L. (2000). *Land Use Management Mapping for the Murray-Darling Basin.* Report to the Murray-Darling Basin Commission. Canberra, ACT: Bureau of Rural Sciences.

Volume 3, Multi-Criteria Analysis Publications, 2005–2014

MCAS-S Development Partnership (2014). *Multi-criteria Analysis Shell for Spatial Decision Support (MCAS-S) User Guide* (version 3.1). Canberra, ACT: Australian Bureau of Agricultural and Resource Economics and Sciences.

Lesslie, R. (2012). Mapping our priorities—innovation in spatial decision support. In P. Figgis, J. Fitzsimons & J. Irving (Eds.), *Innovation for 21st Century Conservation* (pp. 46–153). Sydney, NSW: Australian Committee for IUCN.

Australian Bureau of Agricultural and Resource Economics and Sciences. (2011). *Multi-criteria Analysis Shell for Spatial Decision Support (MCAS-S) User Guide* (version 3). Canberra, ACT: Australian Bureau of Agricultural and Resource Economics and Sciences.

Lesslie, R., Thackway, R., and Smith, J. (2010). *A national-level vegetation assets, states and transitions (VAST) Dataset for Australia* (version 2.0). Canberra, ACT: Bureau of Rural Sciences.

Stenekes, N., Kancans, R., Randall, L., Lesslie, R., Stayner, R., Reeve, I. & Coleman, M. (2010). *Indicators of Community Vulnerability and Adaptive Capacity Across the Murray Darling Basin—A Focus on Irrigation in Agriculture*, Report to the Murray Darling Basin Authority. Canberra, ACT: Bureau of Rural Sciences; Armidale, NSW: Institute for Rural Futures, University of New England.

Cotsell, P., Gale, K., Hajkowicz, S., Lesslie, R., Marshall, N. & Randall, L. (2009). Use of a multiple criteria analysis (MCA) process to inform Reef Rescue regional allocations. In S. Hogan & S. Long (Eds.), *Proceedings of the 2009 Marine and Tropical Sciences Research Facility Annual Conference,* 28–30 April 2009. Reef and Rainforest Research Centre Limited, Cairns, NT.

Lesslie, R., Smith, J. & Rickards, J. (2009). Agricultural land use mapping and decision-support in Australia. In Y. Chen, Y. He, & Q. Yu (Eds.), *Agricultural Land Use and its Effect in APEC Member Economies* (pp. 1–19). APEC Agricultural Technical Cooperation Working Group, Beijing.

Byron, I. & Lesslie, R. (2008). Spatial methodologies for integrating social and biophysical data at regional or catchment scale. In R. j. Aspinall & M. J. Hill (Eds.), *Land Use Change: Science, Policy and Management* (pp. 3–62). Boca Raton, FL: Taylor Francis Group.

Lesslie, R. & Cresswell, H. (2008). Mapping priorities: planning re-vegetation in southern NSW using a new decision-support tool. *Thinking Bush 7,* 30–3.

Lesslie, R. G., Hill, M. J., Hill, P., Cresswell, H. P. & Dawson, S. (2008). The application of a simple spatial multi-criteria analysis shell (MCAS-S) to natural resource management decision making. In C. Pettit, W. Cartwright, I. Bishop, K. Lowell, D. Pullar & D. Duncan (Eds.), *Landscape Analysis and Visualisation* (pp. 3–95). Lecturer Notes in Geoinformation and Cartography Series, Springer, Berlin.

Zhang, K., Liu, N., Fu, Z., Lesslie, R. & Caelli, T. (2007). Discovering prediction model for environmental distribution maps. *11th Pacific-Asia Conference on Knowledge Discovery and Data Mining* (PAKDD'2007). CD-ROM.

Buang, N., Liu, N., Caelli, T., Lesslie, R. & Hill, M. J. (2006). Discovering knowledge from distribution maps using Bayesian networks. *Proceedings of Fifth Australasian Data Mining Conference* (AusDM2006) (pp. 9–74). Sydney.

Hill, M. J., Lesslie, R., Donohue, R., Houlder, P., Holloway, J. & Smith, J. (2006). Multi-criteria assessment of tensions in resource use at continental scale: a proof of concept with Australian Rangelands. *Environmental Management 37*(5), 712–31.

Lesslie, R., Hill, M., Woldendorp, G., Dawson, S. & Smith, J. (2006). *Towards Sustainability for Australia's Rangelands: Analysing the Options.* Canberra, ACT: Bureau of Rural Sciences.

Donohue, R., Hill, M., Holloway, J., Houlder, P., Lesslie, R., Smith, J. & Thackway, R. (2005). *Australia's Rangelands: An Analysis of Natural Resources, Patterns of Use and Community Assets.* Report to the Natural Resource Management Business Unit, Department of Agriculture, Fisheries and Forestry. Canberra, ACT: Bureau of Rural Sciences.

Hill, M. J., Lesslie, R., Barry, A. & Barry, S. (2005). A simple, portable, spatial multi-criteria analysis shell—MCAS-S. In A. Zerger & R. M. Argent (Eds.), *MODSIM 2005 International Congress on Modelling and Simulation.* Modelling and Simulation Society of Australia and New Zealand, Melbourne, December.

www.ingramcontent.com/pod-product-compliance
Lightning Source LLC
Chambersburg PA
CBHW050807270326
41926CB00026B/4596

9 7 8 1 9 2 1 9 3 4 4 1 4